Drones and Journalism

Drones and Journalism explores the increased use of unmanned aerial vehicles, or drones, by the global media for researching and newsgathering purposes. Phil Chamberlain examines the technological development and capabilities of contemporary drone hardware and the future of drone journalism. He also considers the complex place of the media's drone use in relation to international laws, as well as the ethical challenges and issues raised by the practice.

Chapters cover topics including the use of drones in investigative reporting, in reporting of humanitarian crises, and the use of this new technology in more mainstream media, such as film and TV. The book also presents exclusive interviews with drone experts and practitioners and draws on a wide range of disciplines to put the practice into a historical, political and social context.

Professionals and students of Journalism and Media Studies will find this an important critical contribution to these fields, as Phil Chamberlain astutely charts the rise of the reliance on drones by the media worldwide.

Phil Chamberlain is Associate Head of the Department for Broadcast & Journalism at the University of the West of England, UK. He has 20 years' experience as a journalist, and has worked for UK national newspapers on investigative projects. He is the co-author of *Blacklisted: The Secret War between Big Business and Union Activists* (2015).

Drones and Journalism

How the Media is Making Use
of Unmanned Aerial Vehicles

Phil Chamberlain

Routledge
Taylor & Francis Group

LONDON AND NEW YORK

First published 2017
by Routledge
2 Park Square, Milton Park, Abingdon, Oxon OX14 4RN

and by Routledge
711 Third Avenue, New York, NY 10017

Routledge is an imprint of the Taylor & Francis Group, an informa business

© 2017 Phillip Chamberlain

British Library Cataloguing-in-Publication Data
A catalogue record for this book is available from the British Library

Library of Congress Cataloging-in-Publication Data
A catalog record for this book has been requested

ISBN: 978-1-138-66878-2 (hbk)
ISBN: 978-1-315-61847-0 (ebk)

Typeset in Times New Roman
by Apex CoVantage, LLC

Contents

The author would like to thank all of those who took time to be interviewed for the book. Any errors are the authors and not the interviewees. The author would also like to thank his colleagues at the University of the West of England for their support, ideas and encouragement.

Introduction

There were no regular inhabitants of the village, but often people visited it to practice fighting. The village sits on Salisbury Plain, a huge swathe of England used by the British Army since Napoleonic times to train its troops. In 1990 the village had been modelled to look German but as threats changed so had its look.[1] In 2008 when I visited, it was a melange of Iraq and Afghanistan and was hosting a competition.[2]

The object of the competition was to come up with a system to help ground troops enter just such a village and deal with various hostile forces from snipers to improvised explosive devices (IEDS). There were half a dozen teams and each was made up of small companies and a university. They were demonstrating to various men from the ministry how their systems worked in the hope of securing a contract. It was deliberate that universities and small companies had been paired, as the ministry was looking to increase the range of organizations that might bid for contracts. Having studied the problem most of the teams had come up with some kind of drone as a focal point for their efforts.

In a small van packed with equipment and discarded coffee cups, staff from one of the teams were checking on the progress of their various appliances several hundred yards away. One of the university engineers, unshaven, having spent two days living in an army tent and clearly enjoying every minute, talked through what I could see on a monitor. What I could see were some very fuzzy pictures looking over the village. He pointed to a crumpled giant fly on the mud next to the truck. "The weather has been difficult so we've had some losses but it is one of these that is sending these images back," he said. He picked it up and added as an afterthought: "In ten years Sky News will have a fleet of these."

I looked dubiously at the fragile structure and then went back to the monitor. A truck with an anti-aircraft gun on the back was hiding behind a 'mosque'. On the monitor it had been framed by white bars as its outline had triggered an alert from an algorithm which looked to match shapes seen by the drone's camera to a visual dictionary of threats. The engineer had been ambitious in his prediction about the use of drones by the media, but not by much.

Take these headlines from the summer of 2016. An earthquake in Italy kills nearly 300 people and within 24 hours footage shot by drone of the devastation is aired on international news channels (*Guardian* 2016). In Virginia a 65-year-old woman shoots down a drone which has been flying over her house which borders that of film star Robert Duvall (Farivar 2016) while Utah becomes the latest state to give the authorities the power to disable drones in specific circumstances (AP 2016).[3] There is a report of a near miss at Exeter airport between a drone and a passenger aircraft (Oldfield 2016). According to the UK union representing airline pilots, the number of reported incidents involving drone near-misses with planes has increased from 29 in 2015 to 42 by September 2016 (BALPA 2016). An air ambulance warns that the fictional plot in a soap opera about a drone bringing down a plane could come true (Bieber 2016). At a carnival in the English seaside resort of Swanage, the brochure has a map showing the parade route and details for when the Punch and Judy is on. There is also a notice that a drone will be flying overhead to gather publicity footage and who to contact if anyone is concerned.

From the serious to the banal and with a constant murmur of anxiety, the use of civilian drones, and in particular their use for newsgathering, has become increasingly prevalent. At the end of August 2016, the Federal Aviation Authority finally published its rules on the commercial use of small drones (FAA 2016). Matt Waite, from the Drone Journalism Lab at the University of Nebraska, said that as a result:

> It's likely that many hundreds of the eventual thousands of licensed drone pilots will be journalists. Many of them are climbing that first hill this week and taking the test. The second big hill to climb is professionalizing operations in newsrooms.
>
> (Waite 2016)

A week before the new regulations came into effect, CNN unveiled its drone division featuring a dozen craft and two full-time operators (Poynter 2016). We are at a crucial juncture where technology, regulation, corporate interests and personal opportunity make it only a question of how ubiquitous drone use by the media becomes – we're beyond the stage of whether it will happen.

This book will look at the development of the media use of drones – how they are being used to tell stories and who is designing and operating them. It runs from investigative journalists to war photographers, television drama departments and advertising agencies; NGOs to hobbyists. It will also consider the changing regulatory framework which covers such issues as privacy and public interest. We'll look at the technological opportunities and limitations. We'll investigate how those teaching journalism are introducing drones into the classroom. And we'll do this within a critical framework which considers what kind of insight this god-like view from above gives us. This isn't a manual on how to fly a drone and it won't look in detail at their current military use – though that is part of the story. It is about what it means for journalism and society when major media organizations, freelance photographers and citizen reporters have their own eye in the sky.

Some will already have strong feelings just by the use of the word 'drone'. It is a noun that is loaded with meaning. There are, or course, other ways of describing the system, including Unmanned Aerial Vehicle (UAV) or System (UAS) or Remotely Piloted Aircraft (RPA). To varying degrees these are all more accurate descriptions. They acknowledge that there is not just the flying unit but a control unit and human involvement. The debate about terminology is explored later and helps illuminate the developing narrative of drones. For the purposes of this book, the term most commonly used and recognized by those in the media, and this book is focussed on the media in the widest sense, is 'drone' and for ease of understanding that is what will be used in this book.

But first; how did we get from a World War One naval technology to celebrities being buzzed over their homes?

Notes

1 Copehill Down is one of several such villages built for troops to practice urban warfare. One of the officers involved in running such training sessions talks about it here: https://insidedio.blog.gov.uk/2015/07/14/enhancing-urban-ops-training-at-copehill-down/
2 The Ministry of Defence Grand Challenge, which appears to no longer run. Government publicity brochure on the 2008 entrants available here: http://webarchive.nationalarchives.gov.uk/20140410091116/http:/www.science.mod.uk/codex/documents/codex_issue2_gc_supplement.pdf
3 A common reason for introducing such laws is the flying of drones over bushfires, which means that planes carrying water cannot carry out any operations. The laws are not designed for the police to interfere to protect the privacy of residents.

References

AP. (2016) Utah will let authorities disable drones near wildfires. *CBC News.* 13 July 2016. http://www.cbsnews.com/news/utah-will-let-authorities-disable-drones-near-wildfires/
BALPA. (2016) Pilots call for greater safety in the skies at TUC. *BALPA web site.* 14 September 2016. https://www.balpa.org/Media-Centre/Press-Releases/Pilots-call-for-greater-safety-in-the-skies-at-TUC

Bieber, N. (2016) Northants air ambulance issues message to public amid helicopter crash on BBC One's Casualty. *Northants Herald & Post.* 27 August 2016. http://www.northampton-news-hp.co.uk/northants-air-ambulance-issues-message-to-public-amid-helicopter-crash-on-bbc-one-s-casualty/story-29660303-detail/story.html

FAA (2016) The FAA's new drone rules are effective today. *FAA web site.* 29 August 2016. http://www.faa.gov/news/updates/?newsId=86305

Farivar, C. (2016) Woman shoots drone: 'It hovered for a second and I blasted it to smithereens.' *Ars Technica.* 29 August 2016. http://arstechnica.com/tech-policy/2016/08/65-year-old-woman-takes-out-drone-over-her-virginia-property-with-one-shot/?utm_content=buffer88740&utm_medium=social&utm_source=twitter.com&utm_campaign=buffer

Guardian. (2016) Drone footage shows extent of earthquake damage to Amatrice. *The Guardian.* 24 August 2016. https://www.theguardian.com/world/video/2016/aug/24/drone-footage-shows-extent-earthquake-damage-amatrice-italy-aerial-video?CMP=share_btn_tw

Mullin, B. (2016) CNN just launched a new drone division. Here's what they plan to do with it. *Poynter.* 19 August 2016. http://www.poynter.org/2016/cnn-just-launched-a-new-drone-division-heres-what-they-plan-to-do-with-it/427027/

Oldfield, E. (2016) Passenger plane in near-miss with drone close to Westcountry airport. *Express & Echo.* 11 August 2016. http://www.exeterexpressandecho.co.uk/passenger-plane-in-near-miss-with-drone-over-the-westcountry/story-29609775-detail/story.html#qbuMjEMmOmWffheY.99

Waite, M. (2016) Announcing: The drone journalism lab operations manual. *Drone Journalism Lab.* 1 September 2016. http://www.dronejournalismlab.org/

1 The origins of unmanned aerial vehicles

When one of the organizers of the London 2012 Olympics was asked what the opening event would look like, they responded: "It's a media event, so it will look great from the air." As Mark Dorrian observed: "Much valued for its spectacular and entrancing effects, the aerial view is firmly established as a recurrent feature of popular visual culture, media forms and touristic installations" (Dorrian and Pousin 2013: 295). And increasingly supplying that footage are drones. As well as being deployed for films, adverts, dramas, documentaries and news stories, the flying eye is now embedded in our culture. They have Twitter and Tumblr accounts,[1] they are raced with prizes in the tens of thousands of dollars and airing on ESPN (Zaleski 2016), they inspire art festivals[2] and no first-person shooter game is complete without one.[3] Films are made about them[4] and TV comedies use them as plot devices. That silhouette of either the dome-headed elongated Predator or the spider-like quadrocopter is instantly recognizable. Yet the drone occupies an uneasy space. When the American comedy *The Big Bang Theory* had its geeks take ownership of one, they were put in peril when the drone took on a life of its own.[5] Visit a scenic sight in America such as Lake Tahoe and drone iconography is deployed to warn you against flying them. As Rothstein writes:

> Drones have become a singular inflection point of fear, of paranoia, of wonder, of technological wizardy, and of future possibility. No other word would suffice, at this point in history, to refer to this web of concepts, meanings, and esthetics so easily.
>
> (Rothstein 2015: 135–136)

The visual spectacular, the promise of technology and the influence of the military have influenced our reaction to, and use of, the drone.

During World War One, combatants first experimented with radio-controlled aircraft and bombs. Once the conflict ended, research continued in a more haphazard fashion. The American Navy's N-9 seaplane was remotely piloted on 15 September 1924 – unfortunately sinking after a heavy landing. Funding was curtailed but it demonstrated the possibility. The term 'drone' was applied to the development of remote-controlled targets developed in the 1930s. The British Navy had target planes for gunnery practice operating under the name Queen Bee, which made a buzzing noise. The US Navy decided to adopt a similar project and according to Callahan: "The name was likely suggested by NRL's Hoyt Taylor, who was confident that 'to those who know anything about honey bees, the significance of the term will be clear. The drone has one happy flight and dies'" (Callahan 2014: 112). The advent of World War Two gave the research a new impetus. Remotely piloted aircraft were used by America in the Pacific in 1944 and Germany's well-known V-1 and V-2 programmes were part of this trend towards automation. Rothstein identifies the first public announcement of a camera attached to a drone in a press release from the American company Ryan Aeronautics in 1955. He writes: "Training remained a useful end, but reconnaissance was to be the drone's talent, once technology made it possible" (Rothstein 2015: 28).

Surveillance is a fundamental part of what makes a drone useful but it has always been problematic. Looking out of the window of his flat, Winston Smith, the hero of George Orwell's novel *1984*, observes "in the far distance a helicopter skimmed down between the roofs, hovered for an instant like a bluebottle, and darted away again with a curving flight. It was the police patrol, snooping into people's windows" (Orwell 1990: 4). It is reported that Pashtun tribespeople in Pakistan refer to drones as "wasps or mosquitoes due to their sound. This sound has a psychological effect on the people who hear it for days on end" (Rothstein 2015: 131–132). The animalistic imagery of Orwell's surveillance tool is deliberate but it is uncanny how the description also matches the flight of today's drones. Smith was not at work when this observation takes place, where he might expect to be monitored. Instead he was at home and the then shocking implication was that such snooping could not only invade the domestic but it could also come unannounced. Smith might have the pleasure at that moment of watching the watchers but the next time he could be the subject. In 2016, TV celebrity Richard Madeley, wearing little but his underpants, was reported to have chased people down the road after a drone had been flown over his home. The Daily Mail ran an interview with the 19-year-old who had flown the drone over the house (by mistake he said) as he was just testing out a present (Dunn 2016). Madeley had swiftly gone on to Twitter to report the incident and post details of the vehicle which had carried the pilots so they could be apprehended. So in sixty years, the darting bluebottle outside the bedroom window has arrived, but its pilots are teenagers. The details are shared almost immediately with anyone who has access to Twitter rather than pondered alone. Rather than the total surveillance of Orwell, we have Thomas Mathiesen's synopticon: thousands of us observing the few.

Before considering the role of drones in the media currently, it is worth looking into how we got to this stage. Not just a few years back to when the technology broke through, but further back to when we were similarly challenged. We should not assume that we got here through a natural chain of events whose outcome was preordained. As Carolyn Marvin writes in her study on the nineteenth century fascination with electric light:

> We often see it as the process by which our ancestors looked for and gradually discovered us, rather than a succession of distinct social visions, each with its own integrity and concerns. Assuming that the story could only have concluded with ourselves, we have banished from collective memory the variety of options a previous age saw spread before it in the pursuit of its fondest dreams.
>
> (Giddings and Lister 2011: 40)

Not only might these other options help inform current debates, but why particular choices were made, or indeed if they were choices, can reveal the social and economic forces at play.

In 1889 the Eiffel Tower opened to the public and almost a million people rode the 324 metres to the top. As Robert Hughes points out, until then most people "lived entirely at ground level, or within forty feet of it, the height of an ordinary apartment house. Nobody except a few intrepid balloonists had ever risen a thousand feet from the earth." When they arrived at the summit, visitors

> saw what modern travellers take for granted every time they fly – the earth on which we live seen flat, as pattern, from above. As Paris turned its once invisible roofs and the now clear labyrinth of its alleys and streets towards the tourist's eye, becoming a map of itself, a new type of landscape began to seep into popular awareness. It was based on frontality and pattern, rather than on perspective, recession and depth.
>
> (Hughes 1991: 14)

Mapmakers and painters had long imagined the view from above. The vision they had created tended to be contained; everything could be grasped in the one image with the boundaries of the city neatly set out. The perspective was often oblique, looking towards the horizon rather than directly down. In the latter half of the eighteenth century, the technology developed to get people up in balloons, but it had been complicated enough getting aloft without then recording what they saw. When Thomas Baldwin produced his *Airopaidia* in 1786, giving views of what he had seen cruising over Chester, he had carried with him a specially created set of pencils. People had the chance to imagine such a lofty perspective themselves with the rise in panoramas in the eighteenth century, which created huge scenes from above where perspective disappeared.[6] As photographic technology developed, it took until the end of the nineteenth century for it to work in balloons. Felix Tournachon, commonly known as Nadar, the French balloonist and photographer, made his first ascent in 1857 but failed to record anything because of problems with the film. He then perfected the ability to develop his plates while aloft. By the first decade of the twentieth century, the trade press were advertising equipment to film from above whether in planes or balloons and giving tips on how to do this effectively. Teresa Castro describes one such article by French balloonist Andre Prothin:

> [He] was clearly more interested in the conventional panoramic possibilities of such vision, arguing that what distinguished them was their documentary powers, their visibility, their topographic qualities and their evident value for reconnaissance, which is to say their cognitive value. On the horizon of this concept lies a conception of aerial vision that is eminently instrumental and functional, an idea expressed here in relation to its extraordinary expansion of the point of view. In the first decades of the twentieth century, an almost blind and widespread belief in the objectivity of methods of mechanised reproduction – such as photography and cinema – only accentuated in the teleological tendency which saw indexical images obtained from the air as the natural replacement for cartographic images.
>
> (Dorrian and Pousin 2013: 123)

As Barber and Wickstead point out: "Aerial views, we argue, are not always the same. . .. Numerous analyses demonstrate how ways of seeing are historically and culturally situated" (Barber and Wickstead 2010: 237). Yet, the exhilaration felt by Baldwin, Nadar and all those who ascended the Eiffel Tower at these new panorama mirrors the excitement that drone footage can inspire today. Suddenly the technology has closed a gap. There is a possibility of seeing something fresh. There is a belief in the power of this new image to create change. Gynnild describes drones as a "disruptive innovation" that has

> emerged accidentally, but disrupts existing conceptions of journalism and subsequently contributes to the creation of new markets and value networks in addition to reducing human risk taking when covering catastrophic and conflicting events. I also argue that the journalistic hunt for the visual conquering of formerly unwatched realties supports the ongoing transition from a norm-based mindset to a more innovative one among professional journalists.
>
> (Gynnild 2014: 336)

Adam Najberg was a journalist with the *Wall Street Journal* for more than 20 years before going to work with the Chinese drone manufacturer DJI. He said:

> Drones present a major opportunity to tell not just old stories from a new perspective, but a completely new way to tell a story. DJI's drones have flight controllers on board, which makes them stable, even when you take your hands off the controls. Even absent GPS signal, such as inside caves, you are able to go where humans cannot easily go, see things humans cannot easily see and tell stories you could not tell before. No drones, no cave video, no story.
>
> (author interview, February 2016)

According to Kellner, we should not be surprised at this immediate demand for the grand spectacle, as the internet-based economy "deploys spectacle as a means of promotion, reproduction and the circulation and selling of commodities" (Kellner 2003: 1). The final aspect to be considered in this chapter is how drones contribute towards our increased desire for the spectacle. In 2015 the BBC used a drone to film Auschwitz-Birkenau concentration camp.[7] Drones have been flown through firework displays and circled above climbers navigating awe-inspiring peaks. Writing in *Corporate Knights* (which bills itself as the magazine for clean capitalism), journalist Tyler Hamilton gave the BBC film as an example of "the good journalism" that drones can aid and described it offering "perspective and insight into a Nazi death camp 70 years after it was liberated" (Hamilton 2015). The 2 minute 31 second-long film has no narration, only sombre classical music and brief screen text to identify particular structures. It begins by mimicking the journey along the train tracks and then there are various shots along and over buildings and fences. It was filmed during winter, with a dusting of snow everywhere. There are few people or anachronistic elements such as cars visible and the film has a slightly grainy texture. It is undoubtedly haunting, though sitting at the back of my mind was Spielberg's *Schindler's List* as an unwarranted quality comparison. I am not sure, though, that beyond its aesthetics, it gave me any new insight. There is a long-established specialism in the use of aerial imagery for archaeological purposes. Studying the changing contours and colours of the countryside can reveal buried remains from thousands of years ago. I am not sure that any historian of the genocide found new information as a result of this impressively mounted footage. Tyler then goes on to list what else drones might cover:

> How about aerial shots of the Fukushima disaster site, or images in the aftermath of natural disasters or areas battered by war? Traffic reports? Police chases? In terms of sustainability reporting organizations could use drones to capture industrial impacts on nature or companies try to evade environmental regulations.
>
> (Hamilton 2015)

It is instructive that the first reach is for the most spectacular and then at the end it is investigative journalism with the strongest public-interest justification attached. Such prioritizing is what many fear could happen as drones become more widely used. Of course, drones were actually used at Fukushima because they were a much safer way of monitoring damage than sending in emergency personnel. This is likely to be a growth area for commercial drone development. One imagines that in turn media companies will request access to such footage much as they sometimes get dash-cam footage from police cars or other emergency services.

The strategic advantage of being higher than others is made clear by the eponymous hero of Swift's *Gulliver's Travels* when he is transported by the flying island of Laputa. The application of magnets keeps this gigantic stone island cruising above the earth. Should the king in his flying fortress be threatened, he would turn his gaze upon the recalcitrant subjects. The hovering island would block the sun above any rebellion, thus causing crops to fail or rocks to be cast down on those below. "Against which they have due Defence but by creeping into Cellars or Caves, while the Roofs of their Houses are beaten to pieces" (Swift 2003: 159). The appearance of this authority then would be one to dread and submit to rather than be negotiated with. A view from above is about a strategic advantage and, in particular, the power of surveillance, which can be thought of in different ways. It can be the panoptic surveillance of Foucault where one discretely observes the many as a form of control. There is also synoptic surveillance which covers both the voyeuristic, where many people might observe a celebrity, and where the tables are turned on the watched, such as monitoring the police. There is also sousveillance, observation from 'below' such as activists recording what happens at demonstrations. What is interesting about drones, now the technology has developed to allow them to break into the mainstream, is that all these versions of surveillance are being practiced. An analysis by Tremayne and Clark of eight instances of use of drones by the media found that most of them matched synoptic surveillance rather than Foucault's oppressive panoptic. Examples included paparazzi footage of Paris Hilton on a beach, activists flying one over their own demonstration and reporters using them to investigate a government detention centre. Certainly in the last case this seems to address all the possibilities of the technology – giving access to a story that otherwise might not be aired and with a strong public-interest justification. We'll be looking at such uses in more detail later; however Tremayne and Clark warned:

> It is not hard to imagine such cases as corporate media firms begin to use drones more frequently. Crime coverage, for example, is already an area where commercial outlets are likely to adopt a government point of view. Use of drones for media coverage of crime could also follow this pattern.
>
> (Tremayne and Clark 2014: 242)

This warning is echoed by Lyon in his general analysis on surveillance. Written before the advent of drones, it nonetheless provides some useful pointers about the way surveillance is deployed and developed – and by whom. Specifically Lyons highlights: "the asymmetrical relationship between corporate organisation and individual consumer" (Lyon 1994: 150). The drone industry is one marked, as many in the technology sector, by constant updates and refinements. The cycle of updated versions may not be as relentless as in the mobile phone sector, but competing companies and new entrants are continually fighting for market share. Improved cameras, longer flight time, cheaper units all increase the surveillance capabilities and attractiveness to journalists, without necessarily addressing fundamental questions about what the right limits are. As Lyon writes:

> If technological advancement produces a perceived problem then some technological fix – encryption, enhanced security – or legal remedy – data protection or privacy law – can be applied to overcome it. This kind of solution basically accepts the status quo while acknowledging that improvements are always desirable.
>
> (Lyon 1994: 162)

For Deuze we have moved on from the panopticon so that we are part of a mesh of overlapping media:

> Like media, power is everywhere yet nowhere – as it generally seems invisible or hardly noticeable to us. And much like media, such power is omnipresent, always watching (and recording) us in everything we do. As our lives move into media, monitoring has become a mundane, perhaps even desirable, part of social subjectivity.
>
> (Deuze 2012: 107)

The drone is part of that mesh. We might look down at our phone as we record and are also tracked, look up at the security cameras or electronically hand over our details online, and now we must gaze at the sky to see another layer of scrutiny by both the public and private sector. One of the differences with the mobile phone, for instance, is that the current narratives about drones are not uniformly positive. Stories about civilian drones either gaze in awe at some application or wail at the potential for destruction. We either read about how they will make our consumer lives easier, admire footage from an inaccessible point or drones are delivering drugs to prisoners and creating havoc with passenger aircraft. While there are still mobile phone scare stories, generally the technology is accepted as regulated and benign provided the user obeys common-sense rules. If anything scrutiny of the very real problems about surveillance associated with phones is restricted to specialist outlets. Drone coverage by the media has yet to mature and oscillates between moral panics and techno-worship. Hard as though this is to credit now, it used to be that you could file a quick business story as a page filler by announcing that a local company had set up a website. The internet was another place where some people went and, like intrepid foreign correspondents, sent back dispatches from the virtual front line. Now we are integrated, it would be like filing an article to announce that a company had given staff mobile phones. We are currently at that same place in drone coverage. It is enough to announce that a media outlet has secured film shot by a drone to warrant its use – whether it tells us anything or not.

For many journalists the author spoke to in preparation for this book, the potential is clear but there was real concern about the regulatory environment and how also how individual users might 'spoil it' for the professionals. For some it was another free speech battle where government regulations which limited the ability of journalists to do their job should be resisted. There was less a concern about the role corporations might play in how they directed this new technology – though a glance at the history should indicate that is a factor. The battle lines, which will be explored in later chapters, are set out by Holton et al.:

> though UAVs show great opportunity as a tool for journalism, the current regulatory environment poses the largest obstacle to realizing these opportunities. It is important to note, however, that journalist use of UAV is not just a matter of getting cool but otherwise unnecessary aerial footage to bolster the bottom line for media corporations . . . this technology can help address real issues of concern for modern journalism related to access, costs, and even safety. What is more the use of UAVs for journalism raises questions that get to the heart of the role of a free press and free speech in a democracy . . . as such, the stakes in the ongoing fight over the use of UAVs as tools for journalism are potentially greater than many have realized to this point.
>
> (Holton et al. 2014: 642)

Notes

1 See for instance http://dronestagram.tumblr.com/ by James Bridle, which purports to be the social media account of a military drone providing sardonic commentary on its operations.
2 Marco Mancosu lists 25 drone film festivals taking place in 2016 in Europe, America and Australia: https://skytango.com/comprehensive-list-of-drone-film-festivals/ while Mary Ryder explores drone art here: https://www.opendemocracy.net/digitaliberties/mary-ryder/beauty-and-callousness-world-of-drone-art
3 Inevitably there is now a game which puts you in the drone pilot's seat. See https://qz.com/811489/the-drone-strike-game-killbox-plunges-players-into-the-most-dehumanizing-aspect-of-modern-warfare/
4 *Eye in the Sky* starring Helen Mirren released in 2016 is probably the biggest such film to date directly addressing military drones (https://www.theguardian.com/film/2016/apr/17/eye-in-the-sky-review-helen-mirren) but the Internet Movie Database lists more than 200 films which involve them in a plot in some way.
5 You can see a clip here: https://www.youtube.com/watch?v=MPMTwM71olI
6 The Rhineback panorama for instance was eight feet in length.
7 See http://www.bbc.co.uk/news/world-europe-30953301

References

Barber, M. and Wickstead, H. (2010) One immense black spot: Aerial views of London 1784–1918. *The London Journal* Vol. 35(3), 236–254.

Berger, J. (1972) *Ways of seeing.* London: Penguin.

Callahan, A. (2014) Reinventing the drone, reinventing the navy. *New War College Review Summer* 2014, Vol. 67(3), 98–122, 2.

Crary, J. (1992) *Techniques of the observer: On vision and modernity in the nineteenth century* Cambridge: MIT Press.

Deuze, M. (2012) *Media life.* Cambridge: Polity.

Dorrian, M. and Pousin, F., ed. (2013) *Seeing from above: The aerial view in visual culture.* London: IB Taurus.

Dunn, J. (2016) 'Peeping Tom' teenager accused of flying drone over Richard Madeley's house says 'totally bonkers' TV presenter overreacted and chased him in his boxer shorts. *Daily Mail.* 13 April 2016. http://www.dailymail.co.uk/news/article-3538174/Peeping-Tom-teenager-accused-flying-drone-Richard-Madeleys-house-says-totally-bonkers-TV-presenter-overreacted-chased-boxer-shorts.html#ixzz4LLEFXfS1

Giddings, S. and Lister, M., ed. (2011) *The new media and technocultures reader.* London: Routledge.

Hamilton, T. (2015) Drone journalism is coming, like it or not. *Corporate Knights.* 2 February 2015.

Holton, A., Lawson, S. and Love, C. (2014) Unmanned Aerial Vehicles. *Journalism Practice* Vol. 9(5), 634–650.

Hughes, R. (1991) *The shock of the new: Arts and the century of change.* London: Thames and Hudson.

Kellner, D. (2003) *Media spectacle.* London: Routledge.

Lyon, D. (1994) *The electronic eye: The rise of surveillance society.* Cambridge: Polity Press.

Orwell, G. (1990) *1984.* London: Penguin.

Rothstein, A. (2015) *Drone.* London: Bloomsbury.

Swift, J. (2003) *Gulliver's Travels.* London: Penguin.

Tremayne, M. and Clark, A. (2014) New perspectives from the sky. *Digital Journalism* Vol. 2(2), 232–246.

Zaleski, A. (2016) Meet Charpu the drone racing megastar doesn't feel like racing. *Wired.* 9 June 2016. https://www.wired.com/2016/09/meet-charpu-drone-racing-megastar-doesnt-feel-like-racing/

2 Technology and industry come together

On a low-rise industrial estate on the outskirts of Bristol with a constant hum of traffic from the M5 in the background, a group of people are learning to fly drones. There are 11 of them, all men aged between 25 and 55. They've dumped their rucksacks and are sitting around a table with maps spread out in front of them and a computer screen showing a string of letters and numbers. Jim, a discretely worn pair of wings signalling his previous career with the RAF, is showing them how to decipher a Terminal Aerodrame Forecast. This is a time-specific and very localised weather forecast. If you plan to fly your drone, then knowing what the weather might do is crucial. Later on that day the trainees will learn about katabatic and anabatic wind, the Kennziffner Index (measuring solar storms), how to safely transport lithium polymer batteries, the pull effect of flying near buildings and the micro-climates around water. The trainees have all paid more than a thousand pounds for an intensive course with a company called RUSTA so they can earn a Civil Aviation Authority licence to fly a drone. RUSTA, employing former RAF trainers, runs them all round the country and has no problem filling places. Those attending vary from model aeroplane enthusiasts looking for a new challenge to four British Airway employees looking to set up their own company. Gary from Newcastle said: "I expect there to be a lot more regulations in the next few years so I want to make sure I get my certificates now before it is really popular" (author interview, February 2016).

Underlying the different talks is an emphasis throughout on thorough planning and safety. Jim tells them to adopt an attitude of learning from mistakes and be willing to report incidents. The regulations seem to cover endless scenarios from flying displays to using model aircraft without surveillance or why you can't parachute from a craft weighing less than 20 kilograms. There are lots of questions because the regulations sometimes seem unclear or haven't caught up with changes in technology. One wants to know about mounting pyrotechnics on a drone. As the other trainees laugh, Jim gives him a look which has probably been practiced on many an RAF pilot convinced they can defy nature or common sense. Jim, who flies an Inspire drone himself, believes it is likely that licencing will come in, possibly via the police, in an effort to regulate the airspace.

Firms like RUSTA are one example of the growth in the drone industry. In this chapter we will look at the business of drones and how that might influence their use by the media. We'll cover the technological challenges and opportunities and where these are being resolved. It is worth reminding ourselves how we got to this place. Journalism has always been linked with technological change whether the invention of the printing press, the development of photography or the rise in personal computing. The fact that we have ended up with a particular model for journalism was not ordained. As Bijker writes:

> It is sometimes said that we get the politicians we deserve. But if this is true, then we also get the technologies we deserve. Our technologies mirror our societies. They reproduce and embody the complex interplay of professional, technical, economic, and political factors.
>
> (Bijker and Law 1992: 3)

The technology behind the drone is explicitly military. As Rothstein reminds us by comparing aeroplanes and personal computers:

> Drones did not develop across Midwestern and European workshops, in the wind tunnel of a bicycle manufacturing company, in a Silicon Valley start-up, or in the special project division of an international corporation. Drones developed in the American military-industrial complex, and what they look like, how they work, and how they have been used all stem directly from that history.
>
> (Rothstein 2015: 26)

The battle over what to name drones is part of civilian industry's desire to get away from all the negative military associations. A revealing paper by Jackman looked at the rhetoric deployed at drone trade shows, which are increasingly large and lavish celebrations of techno-fantasies where companies call upon investors to create the future they desire (Jackman 2015). Many would cringe at using the word 'drone'. Green writes about how one show had as its WiFi password "dontsaydrones" and Michael Toscano, CEO of the Association of Unmanned Vehicle systems, telling a reporter:

We don't call them drones ... when most people hear the word 'drone', you think military, you think hostile, you think large and you think autonomous. There's a total misconception. And every time the media uses it, you're not portraying good information.

(Green 2015: 243)

Yet it is the military which provides the bulk of the contracts and the research funding. San Diego-based General Atomics had revenues in 1980 of $115 million. Thirty years later that stood at $661 million and the $2.4 billion of products it sold between 2000 and 2010 were dominated by its Predator and Reaper drones purchased by the US military. To protect this mother lode the firm made sure to keep the politicians who signed the cheques sweet with funding of campaign contributions and junkets. "For our size, we possess more significant political capital than you might think," CEO James Blue explained (Benjamin 2013: 34). The big beasts such as Boeing and Lockheed Martin are in the drone military game but this is a sector where specialists firms can pick up some lucrative contracts. Aero-Vironment designs small drones to be launched by soldiers in the field such as the Switchblade, Raven and Hummingbird. The latter was described by Time magazine as one of the best inventions of 2011 and built as a proto-type for the Defense Advanced Research Projects Agency (DARPA). "It's shockingly light – weighing less than one AA battery – but carries, at least during the experimental phase, a shockingly hefty price tag of $4m" (Benjamin 2013: 37).

Militarily, Israel vies with America in being a lead on drone technology but the United Kingdom, China, Pakistan, India and Iran have all invested heavily. Civilian drone investment is strong in the United States, China, United Kingdom and France – the latter home to the successful Parrot company.[1] There are now distinct market segments based on companies targeting particular sectors or technical challenges.[2] Getting people interested in solving the problems of making drones fly further, faster, higher and for longer requires a more subtle appeal than a hulking UAV spewing missiles down on some desert outpost. DARPA, based in Ohio, has been looking at tiny drones which can "flutter down a city street or slip through an open window" and has built a huge urban arena to test them (Benjamin 2013: 46). To help develop next-generation weapons, it launched a crowd-sourcing competition to design a drone that could be carried in a rucksack and when launched hang, vulture-like, for several hours. As Benjamin reports:

Opening up drone design to the public also has the added benefit of normalizing drone warfare among the public. As DARPA's Jim

McComick told the media, "we seek to lower the threshold to entry for hobbyists and citizen scientists," the objective being an "exchange of ideas among a loosely connected international community united through common interests and inspired by innovation and creative thought."

(Benjamin 2013: 47)

The similarity with the Salisbury Plain competition outlined in the introduction is clear.

Respected industry analyst the Teal Group has issued forecasts on the drone industry since 2009 and these show how the military still dominates the market. In 2009 it estimated that UAV spending would almost double over the next decade from the current \$4.4 billion annually to \$8.7 billion by 2019.[3] Over the following years, the doubling in a decade prediction, a Moore's law for drone finance if you like, was repeated with actual production by 2015 worth more than \$6 billion.[4] By then the Teal reports were starting to separate out the commercial and military elements of the market. In its 2014 report Philip Finnegan, Teal Group's director of corporate analysis and one of the authors of the study, said: "Our 2014 UAV study calculates the UAV market at 89% military, 11% civil cumulative for the decade, with the numbers shifting to 86% military and 14% civil by the end of the 10-year forecast." Senior analyst Steve Zaloga added: "The Teal Group study predicts that the US will account for 65% of total worldwide RDT&E spending on UAV technology over the next decade, and about 41% of the procurement."[5]

Its analysis of the civilian market in 2016 predicted that the then \$2.6 billion spent on production would soar to \$10.9 billion in 2025. While investment by Facebook and Google was grabbing the headlines it was other industries which were taking the market share, according to the analysts. It predicted construction will lead the commercial market with all 10 of the largest worldwide construction firms either deploying or experimenting with systems. Behind this would be agriculture. Behind this come energy, insurance and general photography followed by what it called "niche applications" such as delivering humanitarian aid. However, uncertainty about the development of regulations and economic feasibility meant that the Teal Group was unable to make definitive predictions. Civil government use, such as that deployed by emergency services, was seen as a growth area; however, "the consumer drone segment will continue to grow although the explosive increase of the recent past will slow. The continuing increase will be based on easier to use systems with new capabilities and a broader range of suppliers."[6] According to a report from

Price Waterhouse Coopers in May 2016, the potential value of the media and entertainment sector for drones is $8.8 billion – out of a total potential market of $127 billion.[7]

A House of Lords report from 2015 estimated that 150,000 job will be created in Europe in the civilian drone sector by 2050 and that these new jobs would be "spread across manufacturers, operators and the broader supply chain of enabling technologies" (House of Lords 2015: 13). Meanwhile in July 2015 the Civil Aviation Authority had 862 companies registered for small drone work.[8] A search by the author of the UK patent office for patents lodged from January 2010 to July 2016 for drones or unmanned vehicles found 58. Of these, 39 were to do with UAVs, the rest split evenly between land-based vehicles and underwater systems. Of the 58, 38 were published in the last three years and a dozen in the first six months of 2016. The patents came from a range of organizations but the military contractor BAe stood out with eight registered. Others are getting in on the business. Both phone firm Nokia but also heavy machinery firm Caterpillar registered drone patents in 2015.[9] In the United Kingdom, research and development into drones was one of the six largest areas funded by the military between 2008–2011, with some £195 million spent (Parkinson et al. 2013). Much of this would have been carried out by defence companies and universities.

This blizzard of statistics suggests that the civilian drone sector is going to continue to grow but that the media section of it will not carry the same financial clout. Instead the real money will be in sectors such as agriculture and construction. The bulk of research and development will still come as a result of military requirements. Those innovations will have civilian spin-offs and media ones in particular. That relationship does not have to be one-way though. For instance Cousineau writes about how the US military teamed up with sports broadcaster Harris Corporation in an effort to better search millions of minutes of drone footage. Technology developed to quickly choose and rebroadcast clips from football games was applied to the military film stock:

> In addition to making their video library searchable, the US Air force continues to develop the technology in an effort to link embedded text tags with links to other surveillance such as photographs, local maps, cell phone calls, databases, and documents.
>
> (Cousineau 2011: 520)

Media use also packs a bigger punch because people are going to see it and because of its ability to set the political agenda. A report from the

Birmingham Policy Commission on the security impact of drones said that by 2035:

> The technologies relevant to military RPA [Remotely Piloted Aircraft] will have continued to advance including stealth, weight reduction, advanced communications, and the automation of processes such as navigation and manoeuvring. Robotics will have rapidly developed and spread within the civilian economy. With enlightened assistance from government, academia, media, the legal profession, the moral implications of machine autonomy will be clarified and better understood.
>
> (Birmingham Policy Commission 2014: 9)

We should not be surprised at such relationships. Foucault talked of a boomerang effect where technologies tested in colonial wars were then imported to be used domestically.[10] As Graham describes it, a "sprawling transnational industrial complexes that stretch beyond the military and security sectors to span the technology, surveillance and entertainment industries; a wide range of consultants, research labs and corporate universities" (Graham 2010: xvii).

One of the biggest commercial drone companies is Dajiang Innovation Technology – better known as DJI – based in Shenzhen, China.[11] In 2014 it sold 400,000 drones and has seen sales triple or quadruple every year since 2009, giving it a market worth around $8 billion. Founder Frank Wang has a backstory which mirrors Jobs or Gates except Wang was building prototype flight controllers from his dorm at Hong Kong's University of Science and Technology. DJI's Phantom unveiled in 2013 was the first ready-to-fly quadcopter that was robust and affordable. Since then, French manufacturer Parrot, 3D Robotics, GoPro and numerous small companies have entered the market.[12] In an interview with Forbes, Wang said it wasn't competitors which stopped DJI's expansion. "Our main bottleneck for growth right now is the speed at which we come to realize answers to technical puzzles," he said. "You can't be satisfied with the present" (Mac et al. 2015). Adam Najberg is their global director of communications who went to the company after a long career as a reporter. In an interview with the author, Najberg explained how a company such as DJI works with the media:

> I would say that media usage breaks down to a very small, albeit growing part of DJI's business. But in some cases, it is disproportionately higher profile, because media's purpose is to tell stories to a large

audience, and sometimes, the sheer power of the story and footage speaks volumes about the technology.

(author interview, February 2016)

Certainly DJI has been trying to strike partnerships with media outlets and not simply put out a product and see who might pick it up. Najberg said:

> We've worked with *Good Morning America* on storytelling projects several times, including Iceland twice and Vietnam. We have provided equipment and technical advice to the *New York Times* and are in beginning stages of working with other news organizations in Australia and the US to introduce our technology into newsrooms. On its own, the *Times* has already used drones successfully to tell several stories, including one video from the Cook Islands. And, interestingly, one of our flying cameras, the Zenmuse X3, we brought back down to the ground on a smart handle that allows control of the gimbal to stabilize video. That's the Osmo. And it's being used by multiple reporters at the *Wall Street Journal*, *New York Times* and other news organizations to shoot stories where both journalist and subject are on the move. That's an example of how drone technology developed for the air has also contributed to reporting on the ground. And we're just starting to speak with the ICFC about how we can help them help their 90,000 members around the world. Jason Bellini of the *WSJ*, for example, is a power-user of the Osmo. Josh Haner of the *NY Times* regularly uses drones to tell stories.
>
> (author interview, February 2016)

In terms of what comes next, Nasberg is unsure but he cites a crucial part of drone technology – its hackability. He said:

> We hope it's journalists who tell us how they use or want to use our technology. That is what has happened in other vertical sectors. In anticipation of that, we've opened up a lot of APIs onboard our drones. What that means is we're starting to see creation of a robust ecosystem of apps – for mapping, search & rescue, firefighting, for example – that make drone or camera behave a certain way that's best-suited for the task at hand. I can image a media organization telling a story about urban planning in Nigeria, for example, needing to use a mapping program that makes the drone semi-autonomous to capture the photos and detail needed to generate a good map on which

to base the story. Or, perhaps there will be a need for a post-disaster drone-flying pattern that allows media to collect and transmit information without getting in the way of rescue operations. Or maybe even assisting with rescue operations. It's exciting to not really be 100% sure of where developments will go and to work with media to help them and our technology develop alongside each other.

(author interview, February 2016)

One of the criticisms of DJI has been that its software, like Apple, is closed. It's a selling point highlighted by 3D Robotics which trumpets the open source nature of its products. Certainly hackability has been a feature of drone early adopters. Many of those the author spoke to had modified their drones, for instance adding their own cameras, or built them from scratch once they had been confident enough. Regular users know that crashes are an accepted risk of flying drones and that means modular designs where parts can be replaced easily make more sense. The market has expanded to offer relatively complex drones out of the box which adventurous media personnel could use. It has matured much as the camera market so that there are hobby, consumer and professional models and the infrastructure alongside to support people wanting to modify their drone.

Matthew Schroyer, founder of the Society of Professional Drone Journalists, said:

> When you start to look at the drone as just another tool in the kit for next-generation digital storytelling and reporting, and see how it can integrate with other bits such as sensors, code, and virtual reality, things begin to get very interesting. At its most basic, the drone is a platform for any number of sensors. The sensor may be visual, such as an ordinary action camera. But it also could be an infrared sensor, detecting heat and gas emissions. It could be a near-infrared sensor, detecting plant health. And the drone can use all of these sensors to map and digitally reconstruct an environment for data-driven, evidence-based reporting.
>
> (author interview, April 2016)

In terms of technical challenges the situation has moved on remarkably in just a few years. In 2011 it was about soldering and electronic skills. Now, says Schroyer,

> you can walk into a well-stocked electronics store, pick a drone off the shelf, and 30 minutes later you can be flying at 400 feet and taking high-definition aerial video. Drone journalism still requires many hours of 'stick time' to be truly effective, but the trend is towards

easy-to-use, fully-automated flying robots. Object detection and sense-and-avoid are becoming more mainstream. The end-game seems to be taking the human out of the loop. What you're left with, then, is how to handle the data that drones produce. What image overlap is necessary for your photogrammetry software? How much are you going to decimate the mesh? What algorithm are you going to use to detect features, estimate volumes, predict changes? These might be the relevant technical challenges for the drone and data journalist of the future.

(author interview, April 2016)

Some of this work requires skills beyond what an 'ordinary' journalist is likely to possess and instead a flexibility to deal with different types of sensors. Ben Kreimer has looked at putting sensors to monitor environmental conditions into drones. As an article in Poynter explained:

This is inspired in part by an earlier project in Nairobi, where Kreimer made a browser-based 3D reconstruction of a landfill with footage from a drone. That experience prompted him to look into using sensor-equipped drones that could monitor pollutants similar to the kind that emanate from the dump.[13]

Another example, though not from a journalist, comes from Dr John Day and his team at the University of Bristol. They've won funding from Find A Better Way, the charity founded by England and Manchester United legend Sir Bobby Charlton to look at how to improve landmine detection. There are an estimated 110 million active landmines in the world today, most of which are located in less-developed countries. Clearing these mines using current technologies would cost an estimated $30 billion and take over 1,000 years. The Bristol researchers plan to speed up mine detection by flying drones over potential minefields. The drones will obtain high-resolution images that show the terrain and objects visible on the surface clearly. They are developing hyperspectral imaging techniques, which will allow them to obtain a separate image of an area at many different wavelengths or colours of light. Chemicals in landmines can leak out and are then absorbed by vegetation which subsequently changes how they are viewed using particular imaging techniques. The plants are a flag for the presence of mines. These images could detect the effects explosive chemicals have on vegetation as a means of identifying mined areas. Dr Day explained: "Drones taking infrared pictures to map suspected danger zones may provide a quick and safe way to tell if an area is likely to be hazardous."[14] Hyperspectral imaging is an advanced method but the platform chosen is an off-the-shelf drone. When the researchers started they considered building their own drones but there was no need since commercially available units were up to the task. Indeed, as Dr Day explained to the author, sometimes it was enough simply to have the drone fly over an area which locals had previously feared to investigate because of the threat of landmines to show where there might be contamination. If the process was to have any scalability then it also needed to be available to countries which did not have the skills at hand to carry out advanced imaging reconnaissance. The simpler the process the greater the impact it could have. The research at Bristol benefits from the hobby interest shown by one team member in the possibility of drones, indicating that multi-disciplinary approaches which bring together different skill sets can produce the most useful results. The newsrooms of the future may similarly be about bringing together people much as Edison's labs did to solve particular problems.

There are various aspects of drones which are being looked at. One is battery life. The longer a drone can stay airborne the more it can do. Batteries are generally the heaviest part of a drone so the smaller and more efficient they can become the better. This is a key area of research. The development of lithium-ion batteries was one of the enablers that allowed drones to make the leap into general use though they are vulnerable to explosion if not handled correctly. Drones can be either fixed wing or propeller driven and both have benefits. The more propellers fitted the greater the carrying potential, for example. Nonetheless, insects can still navigate turbulent skies with far greater ease than drones which are vulnerable to relatively light winds. This is a key reason for them not gaining greater mainstream use as it only takes mildly inclement weather to turn a flight into a very sudden crash. The composition and manufacturing prowess for rotor design still has some way to go and the failure rate on parts is still far relatively high. The controller interface has increasingly developed to piggyback on mobile phones or tablets, which again has allowed them to be easier to market to consumers. Where some research is looking at is how that interface can be improved so that pilots can perform their job better. There is little feedback other than visually for a pilot and therefore not every sense is being used to ascertain how stable a drone is at any one point. By the time course corrections are made it can be too late. Getting that sensory feedback from the vehicle would improve flying ability. There are huge areas of research looking at sensors and this is where journalism can see many advantages. It is not just about sticking on different types of camera to provide 3D or virtual reality images. There are drones being developed for agriculture that sniff out plant infections. The potential to investigate environmental issues using drones is obvious. The use of algorithms to help examine data or suggest flight plans is also increasing, though the processing power required for the former can be daunting.[15] Another area is the degree of automation. Pilot error is a common cause of crashes and the loss of a phone signal can be catastrophic. A drone developed by Jonathan Aitken and colleagues at the University of Sheffield uses onboard twin cameras to build a three-dimensional view of places and objects it has never seen before. The idea is to give small drones the ability to venture into dangerous areas where humans can't go and make their own decisions on what objects merit closer examination. Knowing where a drone is requires increased GPS capability. Safety is a key aspect, so there is a lot of research around areas such as geofencing where particular sites send out electronic barriers to stop drones approaching.[16] One issue raised is about how drones are controlled in the first place. The already-congested 2.4Ghz that most drone operators currently use will become more of a challenge for operators and this will require engagement by telecoms regulators.[17]

It is worth considering Morozov in *Click Here To Save Everything*, which looks at technical solutions to problems which don't need solving. For Morozov:

> Instead of investigating and scrutinising the motivations for our actions, trying to separate the good ones from the bad, policymakers fixate on giving us the right incentives or removing the option to do the wrong thing altogether. Better safe than sorry as the saying goes.
>
> (Morozov 2014: 343)

There is also the consideration of unforeseen consequences. As social behaviour modifies technology how will that impact be felt? For instance, the explosion in demand for mobile phones has led to a scramble for certain key raw materials in developing countries. Further, Apple has come under fire for the labour practices of factories in China contracted to build its iconic products. Will the development of drones lead to similar environmental and industrial concerns? As Culver writes:

> Technologies throughout history have brought attendant ethical problems, often unanticipated, Advances intended to address specific problems carry unintended consequences. Tenner describes these as "revenge effects" . . . these revenge effects are often directly connected to unforeseen behaviours in the users of technological artefacts. Thus a focus on the social role of technologies and their impact on values, routines, and expectations is critical.
>
> (Culver 2014)

That focus on looking at the wider implications is also highlighted in a paper by Demir et al., which looked broadly at the areas of research around UAVs to chart their status. The authors wrote that "currently it can be said that the UAV domain is in between the phases of technology development and technology acceptance"; that is, that it still needed to mature in key areas. The authors' literature review suggested that particular work still needed to be done on the regulatory, legal, ethical and moral aspects (Demir et al. 2015: 326).

At the moment the future for drones is not yet set. There is still much that might influence how they are used. Jackman detected two dominant rhetorical framings at the trade shows he visited: that of possibility and inevitability. There is a real sense of the industry, and that includes the media, saying that certain developments are inevitable and therefore regulators must act as if things are preordained. It is inevitable drones will be used to deliver packages and therefore sense-and-avoid systems must be developed so cities can be navigated safely (Jackman 2016: 5). Actually the position of drones commercially, including in the media, is not yet set. As Bijker writes on the development of technology and society, an object is not fixed straightaway but subject to negotiation. The idea that it could only develop in one way is the arrogance of determinism. He gives a very useful example of the development of the bike with numerous different versions (the pennyfarthing, tricycle, etc.) brought out to address different issues. There were competing industries and components. It began as a sport before being seen more as a mode of transport. In some European towns laws were passed to protect people from "furious rising" (cycling very fast). At times it had political overtones, being associated with progressive social groups; at others it was linked with the gentry after royalty were pictured using them. During the Boer War the British army commissioned particular versions (Bijke 1994). One can see many similarities with different reactions to drones – how they are used, how they are perceived. Therefore one is likely to see development cul-de-sacs and unexpected uses as drone technology is shaped by and in turn shapes society.

Notes

1 France has taken a lead in 2016 in attracting venture capital for drone companies: http://dronelife.com/2016/05/06/1-country-drone-investment-will-surprise/
2 Useful article looking at different commercial drone companies: https://techcrunch.com/2016/05/14/1319663/
3 See http://tealgroup.com/index.php/about-teal-group-corporation/press-releases/73-teal-predicts-worldwide-uav-market-at-over-62-billion

4 See http://tealgroup.com/index.php/about-teal-group-corporation/press-releases/66-teal-group-predicts-worldwide-uav-market-will-total-89-billion-in-its-2012-uav-market-profile-and-forecast

5 See http://tealgroup.com/index.php/about-teal-group-corporation/press-releases/118-2014-uav-press-release

6 See http://tealgroup.com/index.php/about-teal-group-corporation/press-releases/129-teal-group-predicts-worldwide-civil-uas-production-will-total-65-billion-in-its-2016-uas-market-profile-and-forecast

7 A copy of the PWC report can be found here: http://www.pwc.pl/pl/pdf/clarity-from-above-pwc.pdf, and this report on it illustrates the hyperbole which can be attached to such estimates: https://skytango.com/market-value-of-drone-applications-in-media-entertainment-industry-valued-at-over-8-bn-dollars-says-pwc/

8 Figures reported in http://dronelaw.blogspot.co.uk/2015/07/862-uk-approved-drone-operators.html

9 The construction company spoke about its drone research here: http://www.caterpillar.com/en/news/caterpillarNews/innovation/caterpillar-and-redbird-to-advance-work-site-intelligence-with-drone-analytics.html

10 Foucault gives the example of fingerprinting during the colonial wars in Algeria.

11 There is a good profile on the company here: http://www.uasvision.com/2016/04/25/the-founding-fathers-of-dji/

12 The scramble for a foothold in this market has meant that while there have been lucrative predictions several firms have struggled to replicate the success of DJI. See: http://www.reuters.com/article/us-usa-drones-consumer-analysis-idUSKBN13701Z

13 See http://www.poynter.org/2016/why-2016-could-be-a-breakout-year-for-drone-journalism/390386/

14 For more information see http://www.bristol.ac.uk/news/2016/april/drone-over-old-trafford.html

15 See https://irevolutions.org/2015/10/12/computer-vision-big-data-uavs/ for one example.

16 See https://www.newscientist.com/article/mg22229754.700-one-per-cent/#.VQbNAa1Br00 for one example.

17 See http://www.scl.org/site.aspx?i=ed38095

References

Benjamin, M. (2013) *Drone warfare: Killing by remote control.* London: Verso.

Bijke, W. (1994) *Of Bicycles, Bakelites, and Bulbs: Towards a theory of sociotechnical change.* Cambridge, MA: MIT Press.

Bijker, W. and Law, J., ed. (1992) *Shaping technology/building society: Studies in sociotechnical change.* Boston: MIT.

Birmingham Policy Commission. (2014) *The security impact of drones: Challenges and opportunities for the UK.* Birmingham: Birmingham Policy Commission.

Cousineau, M. (2011) The surveillant stimulation of war: Entertainment and surveillance in the 21st century. *Surveillance and Society* Vol. 8(4), 517–522.

Culver, K. (2014) From battlefield to newsroom: Ethical implications of drone technology in journalism. *Journal of Mass Media Ethics* Vol. 29, 52–64.

Demir, K.; Cicibas, H. and Arica, N. (2015) Unmanned Aerial Vehicle domain: Areas of research. *Defence Science Journal* Vol. 65(4), 319–329.

Graham, S. (2010) *Cities under siege: The new military urbanism.* London: Verso.

House of Lords. (2015) *Civilian use of drones.* European Union Committee, 7th report of session 2014–2015.

Jackman, A. (2016) Rhetoric of possibility and inevitability in commercial drone tradescapes. *Geographica Helvetica* Vol. 71, 1–6.

Mac, R.; Shao, H. and Bi, F. (2015) Bow to your billionaire drone overlord: Frank Wang's quest to put DJI robots into the sky. *Forbes.* 6 May 2015.

3   Flying into the mainstream

The story was a little banal but the prose was breathless. The BBC report on the latest stage of a high speed rail link would not normally have lived long but for the fact that to help tell it the corporation used its own drone to film some of the footage.

Reporter Richard Westcott spoke to the camera which then took off and circled over a railway depot. Accompanying the piece was a backgrounder on the how the BBC had come to use the drone.

One of the reasons the story worked for drone use was because it involved flying over train yards devoid of people and where permission was presumably given by the owners. So for health and safety reasons it was fine. The subject matter might not have been earth-shattering but the potential was clear. Westcott said: "The pictures speak for themselves. You cannot get shots like that with a helicopter, or a steadicam, or a boom, a jib, a dolley" (Westcott 2013).

This foray into drone journalism by the BBC was not something that was rustled up over a few hours. Pilot Owain Rich spent six months getting a licence while he and his team built the six-bladed hexacopter. Actually shooting the package was three to four hours and that included 40 minutes flying time. The BBC machine then got ten minutes charge from each battery. When it came to actually using it Westcott said: "we figured that the best way to milk the machine's capabilities was to use it to surprise the viewer" (Westcott 2013).

Three years later, Owain Rich has been all over the world filming for the BBC using drones. That sense of surprising the audience is still there. Discussing his work for the corporation, Rich said that drones give him the ability to create shots he would not be able to produce with any other piece of equipment: "As a filmmaker I am always trying to find ways of getting a different perspective for the audience, to engage the audience and make them excited. To show them things they might not have seen before."[1]

In this chapter we are going to look at mainstream uses of drones by the media. If this technology is to become a game-changer as some predict (e.g. Gynnild 2014) then it is not necessarily in the unusual areas where this will be seen such as covering major catastrophes. It is how much it is adopted as an everyday piece of kit and whether that is limited to just a few countries. Much as a reporter doesn't think twice about using their mobile phones as a newsgathering tool or a consumer seeing footage obviously shot using one, will the same happen with drones? A report from the Reuters Institute for the Study of Journalism predicted:

> Remotely piloted aircraft will become increasingly available to journalism in the future and news executives and journalists will need to attend to the developments and issues surrounding their use. News organisations will need to make decisions whether it is feasible and desirable to employ them, how they might be effectively used, the ethics of their use in news reporting, and how their use might affect journalistic credibility amongst the public.

> (Goldberg et al. 2013: vi)

One can see in the years since that was published how, if at all, news organizations have used drones. In its very banality the BBC rail story suggested that drones need not be saved for filming erupting volcanoes.[2] A ready deployment would speak much greater volumes about how this technology might become embedded in newsrooms. Indeed for one observer in this area, it's when we're bored that we know they have arrived. Speaking in 2012, Matt Waite, from the University of Nebraska's Drone Journalism Lab, said:

> In 10 years, I think drones for journalism will be boring. I think they'll be like smartphones. They're another tool in the kit. Of course we use UAVs to cover a wildfire. Why wouldn't we? We've been doing it for years. I'm really excited about drones for journalism, I think they'll be a very useful tool for a lot of journalists in the very near future, but I think they're just another tool. Drones will not be writing stories. Drones will not find the humanity in a story. Drones will be a tool for journalists to do what they do best: journalism. And that's all that they are. In 10 years, we'll look back at all this fuss about drones and wonder why we were so worked up.

> (author interview, March 2012)

The media sector still has its behemoths such as News Corp, Sony or Bertelsmann,[3] and these vertically integrated companies produce media across a range of platforms and for a variety of audiences. They support a huge network of other companies and individuals producing specialist items or providing particular expertise. As people move within this sector they take their skills and experience with them and also pick up new ones. Therefore they will draw in drone use among advertising and feature films as well as traditional news media to give a better idea of how drones sit within the ecology of the media industry. Sports coverage is one area with skiing and other action sports ripe for drone coverage.[4] Factual shows are signing contracts with companies to provide drone footage. West Yorkshire-based Aerialrepublic.com won the right to film for BBC's Top Gear that involved travelling to Morocco and Snowdonia National Park in Wales.[5] The six-strong team at the company have worked for a number of media organizations. Advertising companies have been looking not just at creating ads shot by drones,[6] though that is increasingly commonplace, but in using the object itself. So that might be collecting data from phones to send messages via Bluetooth, using them as visual hoardings or skywriting with them to promote films.[7] If one of the drivers for drone use in journalism is cost, that isn't such a factor with feature films where hiring a helicopter for a day is dwarfed by the fee for the on-screen talent. Nonetheless, drone usage has grown enormously even if, as Hennessy points out,

> as it was illegal to film using a drone in the USA up to 2014, most movies still don't list the use of drones on IMDB. In the future, it will become clearer which movies have used drones to film certain scenes instead of cranes or helicopters.
>
> (Hennessy 2016)

As one might expect, action films are the principal user of drones. Bond movies, superhero capers and general mayhem movies have all deployed drones. Among the companies which have specialized in this area are Freefly Cinema and Helicopter Film Services. Operators repeatedly cite the fact that they can fly lower than a helicopter as one of the attractions for drones. Such footage often comes during chase sequences, which inevitably also fill the trailer, so that distinctive drone movement is something which film-goers are increasingly aware of. A genre of films solely shot with drones is developing with a raft of festivals. According to drone director Nikolas Pueringer:

> I think filmmaking as a profession shares a lot with drones in that it is both a very technical profession aligned with something that has a lot of expression and artistry and craftsmanship in it, so the same learning curve applied to drones.
>
> (Watercutter 2015)

As the filmmaking example shows, one particular restriction on drone development is a regulatory one. This will be covered in more detail in the legal and ethical chapter but it is worth giving a brief idea now because it provides context for how they are currently used in the industry. In the United Kingdom, for non-commercial activities, you must keep your drone below 400 feet; at least 50 metres from buildings, people or vehicles; not over congested areas; and always within line of site.[8] These rules are set and enforced by the Civil Aviation Authority (CAA) and they class a drone as being smaller than 20 kilograms. For commercial activities, pilots need to be aware of the Air Navigation Order. This sets out standing rules on how you can use drones and as long as you follow these you don't need to seek permission from the CAA.[9] The order says that a craft with a camera can't be flown over or within 150 metres of any congested area; over or within 150 metres of an organized open-air assembly of more than 1,000 persons; within 50 metres of any vessel, vehicle or structure which is not under the control of the person in charge of the aircraft; or within 50 metres of any person. There will also be restricted airspace around airports and other sensitive areas. Further, anyone who wants to fly a drone commercially needs to get a Permission from the CAA to prove they are competent – essentially a drone pilot's licence.[10] Then, if they want to fly somewhere which would breach the boundaries of the Air Navigation Order, they need to apply to the CAA for a specific Permission for that flight. That would involve carrying out a risk assessment and developing a flight plan, among other requirements.[11] Setting aside questions about what constitutes commercial operations (e.g. if you flew a drone for non-commercial purpose and then accidentally obtained footage which you sold), one can see that this regulatory set-up is the antithesis of what breaking news reporters require. The United Kingdom, unlike America, doesn't require people to register as drone users, so the regulatory burden falls on each individual flight. It can mean that the range of possible flights is much greater but it does make it cumbersome. If every time you needed to jump into a car to drive to the scene of an incident you had to file a drive plan, that would probably go a long way to easing congestion on the roads but also limit the number of reporters who would be sent out on jobs. Instead our driver's licence and the driving test we pass announce that we have basic competency and can be identified and held accountable if necessary. The comparison with motor vehicles is instructive. In 1900 there were an estimated 700 to 800 motor vehicles on UK roads and you did not need to get a licence nor to pass a test.[12] The licence was introduced in 1903, the Highway Code in 1931 and the driving test in 1935. At that point there were well over two million vehicles on the road and more than 7,000 people were dying annually from road accidents.[13] The licencing approach is what has been adopted in America (and some other countries). The Federal Aviation

Authority's Part 107 regulations, which came into force in the summer of 2016, require commercial operators to get a licence to prove their competence and then put in place some standing restrictions, much as in the United Kingdom.[14] Those restrictions include no flying above 400 feet, no night flying, no flying over people, no flying without permission in restricted airspace and always in the line of sight. Some of these restrictions have to be tested (such as the no flying over people) and some may lead to localised agreements such as permissions in low-risk areas which have restricted airspace. Matt Waite said: "Drone journalism will be legally possible in any newsroom in the United States. That's not to say it will be easy, but it will be legally possible in ways that it has never been before."[15] So it remains difficult in America, especially as there are numerous state laws to add layers of complexity, and even more so in the United Kingdom. In addition, as outlined in chapter one, stories about commercial drones are often ones about intrusion or safety concerns. They are about flying drugs into prison, snooping on people in their houses or buzzing aircraft. The perception of drones, even if people are seeing remarkable footage, can be mixed. Steven Flynn is an experienced drone operator in Ireland who said that initially he received a very positive response from the public when he used drones. "In three years this has completely changed to the point of sometimes hostility, sometime oblivious and sometimes aggressive engagement from landowners and the public. . . .. And I know this cannot continue."[16]

According to Tremayne and Clark (2014) between 2010 and 2012 there were eight instances of drone technology being put to use for journalistic purposes. The pair searched databases to identify stories and came back with thousands of hits but only a tiny proportion featured drones as a news-gathering tool. Tremayne and Clark then employed discourse analysis to draw out themes from the stories. The issues were anti-authoritarianism, immediacy and cost, safety, practical limitations, ethics, privacy and the law. "The eight cases identified raise a host of legal, ethical and moral questions which were raised in this report" (Tremayne and Clark 2014: 242). One interesting thing about the eight was the global spread. That ranged from America to Australia and Poland. It was an indicator that this was a technology which was emerging in different places at the same time. While the spread was still in developed countries this was not a Silicon Valley fad.

The BBC has since been joined by other large news corporations in the drone business. They tend to take their time but when they make the jump it tends to create splash. In August 2016, CNN announced the launch of CNN AIR – a division which will specialize in aerial reporting. The broadcaster had spent more than a year putting the unit together and follows a research partnership with the Georgia Tech Research Institute (GTRI) and, separately, with the Federal Aviation Administration. Greg Agvent, senior director for news, told Poynter: "We realized that CNN was in aviation all of a sudden" (Mullin 2016). CNN had been dabbling with drone use for its stories. They'd provided general views as a backdrop to political debates as well as on features about the fiftieth anniversary of the civil rights march on the Edmund Pettus Bridge in Selma, Alabama and coverage of the anniversary of Hurricane Katrina. The company's drone fleet will eventually include a mix of different aircraft from helicopters to fixed-wing planes. The executive identified three different ways he saw drones contributing to CNN: Better production values, greater context and understanding, and improved documentary storytelling. The visual power of drone footage makes it obvious that production values can be improved. Among the stories which CNN has deployed a drone on was a piece on homes in Flint, Michigan affected by lead poisoning in the water. According to Agvent this was an example where drones added important context to a story. They needed to show that it was thousands of homes affected, so the drone started with the reporter in one house:

> The drone was just 10 feet in front of him. And it moved as he moved – outside the door of that individual home, and then it rose above as he went through the rest of his standup to reveal that it's not just one house.[17]

In terms of documentary, CNN has a new programme called 'Great Big Story' which has dramatic footage at its core. "We're able to use the drone to capture things that you simply cannot capture from a helicopter, which would create that much more noise and cost you that much more money," said Agvent. The news corporation's legal department is geared up to help producers who want to make particular programmes. Using a familiar turn of phrase, Agvent said: "CNN's been around for 35 years. We've got a fairly robust system to handle ethical and privacy questions on a daily basis, regardless of what the technology we're using is. This is another tool in our toolbox." It is a phrase used by many journalists the author has spoken to or read about when discussing their use of drones. As Ben Kreimer put it: "I think they are a great tool for journalists to use. They are not going to change journalism, but they're part of a new set of tools that journalists have at their disposal for storytelling."[18] Part of this may be sending out a message of reassurance to audiences and regulators. If it is just another tool, that doesn't make it sound like such a threat. If it is only an incremental change then the current regulatory system for journalists doesn't need a change either. Indeed if regulations treat drones as special and not 'just another tool' then they are based on incorrect assumptions and need to be altered. Many of those willing to talk about drone journalism are advocates who have long laboured for general acceptance of the technology. It is natural they should wish to evangelize and, through experience, persuasion may work best if a change is seen as incremental, such as adopting a new piece of software or getting a different type of phone.

However, that sense of the special can be seen time and again when drone footage is used. An analysis by BBC drone pilot Neil Paton on where drones appeared on news stories found that they tended to provide the opening and closing footage and then a linking piece of footage in the middle. This was roughly the same for any standard 3 minute 30 second or 5 minute 30 second package. It suggests that the attention grabbing nature of the footage is placed where it can have the greatest impact on viewers. The question is whether that format then sends producers out to find footage to fit that construction or if drone footage is still so new that it is enough to hang a story on.[19] In March 2015 the BBC posted a story of drone footage from the Crossrail Tunnel project in London. The text accompanying the film said: "It took four months of negotiating with the Crossrail safety team, but eventually they agreed that a BBC trained pilot, Jon Bontoft, could go down with our transport team, Richard Westcott and Jonathan Sumberg." [20] It is perfectly usual journalistic practice to highlight an exclusive and also common practice to mention early on if it has taken a particularly long or tortuous route to get published. This helps increase the currency of the story. In also matches Paton's analysis that online stories that feature drones tend to be saturated with the footage and all types of stories open with images from the flying camera.

While CNN was forming research partnerships and integrating its drone operation into its corporate structure, agile freelancers have been filling in the gaps a lot quicker. Justin Gong from Sydney was interviewed by ABC in 2012 as he flew his drone around his home town. The son of a retired Chinese air force pilot, he moved to the Australian city to work as a freelance cameraman. He ended up being the Australian director of China-based XAircraft. Gong said:

> I think this stuff is definitely going to change how we do video and photography. It is probably going to take over the helicopter jobs from the emergency news, for example bushfires, floods, earthquakes and also for the evaluation of disaster afterward.

His drones had been giving Chinese viewers a bird's eye view of Australia, as Justin had been shooting news reports for China's Phoenix TV network (Corcoran 2012).

So there has been a mix of large organizations covering specific stories and individuals testing out the boundaries. One gets a sense that such organizations are happy for this to continue for a while as it means that the technical and legal challenges can be addressed by others. The other aspect which will concern organizations is the regulatory environment. This will be looked at in more detail in a later chapter. However, news companies will have noted isolated but worrying incidents of freelance drone journalists being arrested, even if in some cases they seem not to have breached any law.[21] It is worthwhile noting that CNN's announcement of its drone division included details that its legal department was involved, as was the Federal Aviation Authority. It appears the broadcaster is not looking to be transgressive in its use of drones but clearly within the bounds set by the regulator.

According to the Reuters study, journalistic use of drones

> raise[s] a variety of professional and legal concerns, and probably many not currently imaginable, which will influence how, and the purposes which, they are used. They do provide useful functions for news organisations but can have public opinion and legal consequences so news executives will need to consider their uses carefully so as not to negate the benefits they provide.
>
> (Goldberg et al. 2013: 26–27)

One person who has had to consider their use carefully is Joe Myerscough. He is a producer at the independent television production company Windfall Films in the United Kingdom. In 2016 Channel Four aired their two-part programme *Hidden Britain by Drone*, which was based on the premise of drone footage revealing previously unseen parts of the United Kingdom, such as billionaires' private estates and forbidden military zones. It aired at 8pm on consecutive Sundays and was fronted by Sir Tony Robinson, who had started out as a comedian but then made his name leading the archaeology programme *Time Team*. Aerial photography has contributed much to archaeology, so it was appropriate that Robinson should front this series.

Jonathan Holmes reviewed the first programme in the *Radio Times* and his thoughts pick up many of the narratives we are familiar with:

> We're more used to hearing about drones in connection with bombing runs in the Middle East, but now they're invading the stately homes of Britain, and Sir Tony Robinson is the man with his hand on the controls. In this high-concept show, the aerial cameras are used to give a bird's-eye view of sites usually inaccessible to the public, from the eerie fake villages constructed by the Ministry of Defence to practice

invasions – tanks trundling down cardboard streets – to the homes of Russian billionaires, to Winston Churchill's bolthole in case the war went south. The stories are fascinating and, while they could probably have been told without sending up a remote controlled quadcopter, the show is proof that modern technology can also help explore our past.[22]

In an interview with the author, Myerscough explains what it takes to get such a programme off the ground (author interview, July 2016):

Q: How many drones and drone operators did you use – and how did you recruit them?

A: When we started the production we decided to use a single drone company throughout but if you look at the final credits there are 13 pilots and camera operator credited, so it's fair to say our plan didn't really work out. The problem was that we wanted to use the best people, and they tended to get booked up quickly. Because access to our locations was hard to secure we were constantly changing our filming schedule so it was quite a juggling act to find a pilot who was free on the same day as our location would be free, when it wasn't too windy and when we weren't doing anything else.

The other director on the project, Eoin O'Shea, had worked with a drone pilot in the past so we brought him on board and he recommended a number of other pilots who we also used. Then I spent a long time watching drone company showreels, trying to find people who were pushing the envelope a bit. I got to a point where I could tell quite quickly in a phone call whether a company would be a good fit for the production or not. We did need highly skilled operators but attitude was arguably more important. Some operators like to just go off and do their own thing and come back at the end of the day with pretty footage, where we wanted people who would become part of the crew and be open to taking direction. Other operators were very used to shoots where everything revolved around the drone, so it was a shock to the system when action was happening that we couldn't control and they just had to do their best to keep up.

Q: What would you have liked to have filmed but couldn't – and why?

A: One of the strangest things hidden in the British countryside is Hamilton Palace. It's been under construction by property baron and friend of Robert Mugabe, Nicholas van Hoogstraten, on his private estate in East Sussex for the past 31 years but is still unfinished. The estimated cost of the project is over £40 million and when finished it will be the largest one bedroom house in the country. When van Hoogstraten dies it will become his mausoleum. It looks like a Dubai hotel, but in a field in Sussex.

After a great deal of work, we managed to make contact with van Hoog-straten and he eventually agreed to let us fly a drone around the property, but then his son stepped in and asked us not to. We discussed the legality of flying over the site without permission, but none of the neighbours would give us permission to fly from their land and it would have been difficult to guarantee that the flight met safety requirements. We could have flown a helicopter over the site, but that wasn't really the point of the show!

Q: Generally, what do drones offer that adds to a story?

A: Drones are amazingly versatile, it's like a having a helicopter and a jib and a steadicam with you wherever you go and the shots you can get are amazing. But for 'Hidden Britain by Drone' I think they gave us something more, they gave us a distinctive look and feel for the show. Where a direc-tor might choose to use shaky handheld cameras to give a sense of realism in a documentary, by using drones we had a visual way of conveying the sense that we were trespassing, or at least going somewhere you wouldn't normally be allowed to go. By literally flying our drones over walls and fences the whole show ended up feeling sneakier, more covert.

In the future, I'd like to see drones used to tell stories more, to make connections between people and places that would be difficult from the ground and to help explain complex systems.

We all know drone shots of the countryside can be beautiful; one of our stories was set in the Outer Hebrides and the material we gathered there looked stunning, but I think their greatest potential is to show sys-tems at work. When you have a complicated site, like the London Gate-way container port, they're the perfect way to show how everything fits together. Being able to get up high, take in a scene and then fly in to see certain aspects, certain processes and make connections is invaluable. We could follow a single container on its journey through the port in a way we couldn't have captured with cameras on the ground.

The *Hidden Britain* series demonstrated what logistically was required in order to get a prime time television programme on air. There is a law of diminishing returns with regards to such programmes which are hooked on the fact that footage is shot by a drone. At some point public interest will not be excited by just another aerial shot. The lessons from *Hidden Britain* are likely to be picked up by programme makers looking to use drones to add information rather than be the focal point.

One aspect of drone footage which does stand alone is when used by photographers for still images as part of creating a portfolio. Slideshows are a common feature for online media outlets. They might be just the best particular images of the day or grouped around a particular theme. Increasingly drones are that theme or are adding to the portfolio available to picture editors.[23] Or, much as with Paton's analysis of where drone footage appears in news packages, they are used as a way to entice the reader into a story. When the *New York Times* did a piece on the city's use of mass graves, a drone provided the stunning aerial shot above the bleak island to set the scene (Bernstein 2016). As long-time drone journalism advocate and practitioner Ben Kreimer points out, drones are best suited to stories such as natural disasters, large protests, riots and traffic monitoring: "Stories that take place indoors, or stories that, that lack a spatial element will not benefit from drones. Drones are there to add perspective to stories that have a very important spatial element to them, or that take place outdoors" (Norris 2016).

In some ways, developments outside of the United Kingdom, America and other developed nations are more interesting. In July 2016, South Africa hosted its first Drone Camp.[24] It was aimed not just at journalists but activists as well and funded by various philanthropic organizations. There were hands-on sessions to teach people to fly drones and talks from experts such as Prof Paul Ecclestone from the University of Central Lancashire and Buzzfeed's Ben Kreimer. All participants got to take their own 'starter drone' home at the end of the course to help kickstart experimentation in their organizations, and there was a prize for the best drone journalism story idea to come out of the event.

Siyabonga Africa, from the South African Broadcasting Corporation, said: "Drone journalism has provided a new lens with which we can see the world and report on it" (quoted in Olstad 2016).

ICFJ Knight Fellow Chris Roper was one of the organizers. In an interview with the International Journalists Network, he said that the idea was to give a taste of what might be possible with limited resources at an early stage in the technology's development. "I want them to play a role in determining what drone journalism actually is, rather than leave it to non-journalists who happen to be early adopters" (quoted in Olstad 2016).

An example of the kind of work that drones offer is from American photographer Johnny Miller who lives in Cape Town. His series 'Unequal Scenes'[25] uses a drone to capture urban images that reveal the disparities left behind as a result of apartheid. The proximity of slums and golf courses is revealed by the flying eye far better than a ground-level perspective offers. Miller uses maps generated by race, language and income to identify prospective areas and then follows up with satellite images to narrow his target area. One reason is that he only has 12 minutes of battery time for his quadcopter so he can't afford to go on fishing expeditions. Interestingly Miller says that the pictures are not unique – they could be found by using Google

Earth. Clearly, though, his curation, framing and image quality come together to create something greater than the sum of a satellite image.[26]

One of the pioneering people in drone journalism in Africa is Dickens Olewe, who finished a journalism fellowship at Stanford in 2016 and previously set up African Skycam thanks to funding from the African News Innovation Challenge. Skycam was inspired by the use of drones was to cover flooding in western Kenya. It was a deliberate attempt to counter the narrative of the official media who were being helicoptered in on official tours while at the same time trying to question state relief in the area. Drones not only gave a new perspective but were low cost and safer, which meant journalists did not have to risk themselves or their expensive kit to cover the story. As such it is a 'tool' which works well in countries without the same infrastructure as the United States or United Kingdom. Olewe predicts: "African governments are likely to ban recreational operation but license commercial operators. The reason is that they seem to be receptive to a business/research or humanitarian use cases – which they can easily regulate – than open up the skies for everyone when they have great concern about drones being used in ways that they think could threaten country's security." (author interview, April 2016). Nonetheless he has high hopes for the future and in particular sees virtual and augmented reality as well as sensor and data journalism as growth areas. "That kind of story-telling is going to get massive. There is amazing potential." With that growth, though, comes a fear and it is one which sits at the back of the mind of every drone evangelist: "I think you are likely to see more drones and with that the likelihood of something awful happening. It worries me that somebody is going to do something stupid" (author interview, April 2016).

Notes

1 Interview with Rich including footage he has shot here: http://www.bbc.co.uk/news/uk-35712471
2 For just such a film see https://www.wired.com/2014/10/drone-video-iceland-eruption-bardarbunga-volcano/
3 A list of the world's top 100 media companies can be found here: http://www.mediadb.eu/en.html
4 See for instance footage shot by Teton Gravity: http://www.tetongravity.com/video/more-sports/through-the-lens-tgr-drone-cinematography
5 See https://www.insidermedia.com/insider/yorkshire/top-gear-contract-for-aerial-videography-firm
6 There is a growing market for using drones to create marketing films to show off those panoramic vistas. For the individual bored with the usual holiday video companies are now offering drone footage of their trips, see: http://www.nytimes.com/2016/12/28/travel/drone-new-aerial-vacation-photographer.html?_r=0

7 A good round-up of advertising uses can be seen here: https://skytango.com/how-drones-are-changing-the-marketing-industry/
8 A helpful guide from the Civil Aviation Authority can be found here: http://publicapps.caa.co.uk/docs/33/CAP1202droneawareNov15.pdf
9 http://www.caa.co.uk/Commercial-industry/Aircraft/Unmanned-aircraft/Unmanned-Aircraft/
10 http://www.caa.co.uk/Commercial-industry/Aircraft/Unmanned-aircraft/Apply-for-a-permission-to-fly-drones-for-commercial-work/
11 http://www.caa.co.uk/Commercial-industry/Aircraft/Unmanned-aircraft/Guidance-on-operating-permissions-for-drones/
12 Details on motoring history from: http://www.nationalmotormuseum.org.uk/motoring_firsts
13 Motoring regulation details from: https://www.gov.uk/government/publications/history-of-road-safety-and-the-driving-test/history-of-road-safety-the-highway-code-and-the-driving-test
14 Details of the Part 107 Regulations can be found here: http://www.faa.gov/uas/media/Part_107_Summary.pdf
15 Writing in his blog here: http://www.dronejournalismlab.org/post/146262852202/the-faas-drone-rules-are-here-what-does-it-mean
16 Interview with *Irish Technical News* here: http://irishtechnews.net/ITN3/using-drones-for-mobile-journalism-insights-from-steven-flynn-skytango-speaker-at-mojo-april-2930/
17 You can see the footage here: http://edition.cnn.com/videos/us/2016/01/22/flint-water-crisis-lead-gupta-dnt-ac.cnn/video/playlists/flint-michigan-water-cris/
18 Quoted here: http://www.fipp.com/news/features/media-technologist-ben-kreimer-on-storytelling-from-the-air#sthash.EiljOfBS.dpuf
19 Brief details on the Paton findings are available here: https://onlinejournalismblog.com/2016/02/11/what-journalists-need-to-know-to-fly-a-drone/
20 Story can be found here: http://www.bbc.co.uk/news/business-31820033
21 For instance a film in Iran had its drone shot down, see: http://www.ibtimes.com/iran-shoots-down-drone-belonging-film-crew-2465205
22 Review can be found here: http://www.radiotimes.com/tv-programme/e/d4qzcm/hidden-britain-by-drone - series-1-episode-1
23 See this selection of photos by Stephen Zirwes, in which he takes views looking straight down on to beaches and swimming pools: http://www.curbed.com/2016/5/27/11801102/drone-photography-stephan-zirwes-cannes-opiom-gallery
24 See http://dronecamp.codeforafrica.org/ for details
25 See http://unequalscenes.com/
26 A useful article by Meg Miller on this project can be found here: https://www.fastcodesign.com/3060888/exposure/drone-photography-reveals-how-apartheid-shaped-south-african-cities/11

References

Bernstein, N. (2016) Unearthing the secrets of New York's mass graves. *New York Times*. 15 May 2016. http://www.nytimes.com/interactive/2016/05/15/nyregion/new-york-mass-graves-hart-island.html?_r=1

Corcoran, M. (2012) Drone journalism takes off. *ABC*. 21 February 2012. http://www.abc.net.au/news/2012-02-21/drone-journalism-takes-off/3840616

Goldberg, D., Corcoran, M. and Picard, R. (2013) Remotely piloted aircraft systems & journalism: Opportunities and challenges of drones in news gathering. *Reuters Institute for the Study of Journalism*. June 2013.

Gynnild, A. (2014) The Robot eye witness: Extending visual journalism though drone surveillance. *Digital Journalism* Vol. 2(3), 334–343.

Hennessy, U. (2016) Drones in movies: 7 Hollywood movies filmed with drones. *Skytango*. 18 February 2016. https://skytango.com/drones-in-movies-7-hollywood-movies-filmed-with-drones/

Mullin, B. (2016) CNN just launched a new drone division. Here's what they plan to do with it. *Poynter*. 19 August 2016. http://www.poynter.org/2016/cnn-just-launched-a-new-drone-division-heres-what-they-plan-to-do-with-it/427027/

Norris, A. (2016) Media technologist Ben Kreimer on story telling from the air. *Fipp*. 12 September 2016. http://www.fipp.com/news/features/media-technologist-ben-kreimer-on-storytelling-from-the-air#sthash.EiljOfBS.dpuf

Olstad, S. (2016) First-ever drone journalism bootcamp trains South African reporters. *International Journalist's Network*. 1 August 2016. https://ijnet.org/en/blog/first-ever-drone-journalism-bootcamp-trains-south-african-reporters

Tremayne, M. and Clark, A. (2014) New perspectives from the sky. *Digital Journalism* Vol. 2(2), 232–246.

Watercutter, A. (2015) Drones are about to change how directors make movies. *Wired*. 6 March 2015. https://www.wired.com/2015/03/drone-filmmaking/

Westcott, R. (2013) Hexacopter changes the way TV reporters work. *BBC News*. 29 October 2013. http://www.bbc.co.uk/news/business-24712136

4 Covering crisis

The typhoon caused such devastation that the reporter delivers his piece to camera standing slightly awkwardly on the remains of various buildings, as there appears to be no streets. Then we notice that the camera is not static but moving towards him. It gently goes over his head and slopes up several hundred metres, revealing the city's full and utter destruction. A rescue worker and his dog can be seen carefully stepping over the rubble. The worker lifts his hand as if to call for quiet to hear a trapped cry for help. CNN's Karl Penhault continues his narrative as various drone footage is spliced together. One has a fish-eye lens that graphically details the breadth of destruction. Another skims over the water (Penhault 2013).

The city of Tacloban in the Philippines was just one of those ruined by Typhoon Haiyan in 2013. It also saw two international broadcasters deploy drones to try and tell the story of the storm. One was CNN, which used footage from several drones including that of British cameraman Lewis Whyld. The other was the Dutch broadcaster NOS and was shot by cameraman Eric Feijten. NOS had had a drone since that summer but this was the first time it had been deployed for such a story. Feijten said that one of his concerns was that the drone not be seen as a "gimmick" especially with such a serious story. "I think the use of flycams has to add something to the story and the way it is told, like it did in the Philippines," he said. "I usually rest my camera on my shoulder or use a stand or tripod. Using a drone adds another dimension" (Dupuy 2013). According to the *New York Times*, Whyld's efforts had a direct impact:

> Mr. Whyld did not want to beg for a ride on a military helicopter, taking the space of much-needed aid. So he launched a drone into the skies above the city. In addition to shots that showed the scale of the damage, broadcast by CNN recently, his drone discovered two bodies that were later recovered by the authorities, he said in an interview. "The

newspaper was for still images," said Mr. Whyld, who builds his own drones, "but the Internet is for this."

(Kaufman and Somaiya 2013)

As Lisa Dupuy points out in her analysis of the drone coverage: "Although this new technology provides new angles to the telling of news stories, the approach taken by many of the journalists curating these images has not been very different from the other, 'traditional' coverage." She said that reporters focussed on the most severely hit island and city, and after a fortnight interest had moved elsewhere:

> This fluctuation of attention is simply part of the journalistic "ecosystem" that forms around such disaster zones – but indeed, it would be very interesting for these flying cameras to go back in a couple of months or in a year's time, to once again record what traces the typhoon left behind.
>
> (Dupuy 2013)

Actually, and unusually, that is exactly what Lewis Whyld did but this time with the *Daily Telegraph*. "The drone has allowed us to retrace our exact steps six month later," he said. "Previously with a static camera on the ground you wouldn't have been able to do that" (Whyld and Phillips 2014). After 10 days and 20 hours of flight time, Whyld was able to create a split screen film which compared the immediate aftermath with the rebuilding that had occurred:

> There are signs up and everybody knows the rule that you can't build near the sea but it is blindingly obvious that they are doing that. The first thing I saw was that everyone is rebuilding exactly where they were living previously. At the moment it is really up in the air for a lot of families.

That disasters should attract drone images should not be a surprise. As Sontag said: "Ever since cameras were invented in 1839, photography has kept company with death" (Sontag 2003: 12). Baudelaire wrote mournfully in the early 1860s:

> It is impossible to glance through any newspaper, no matter what the day, the month or the year, without finding on every line the most frightful traces of human perversity . . . Wars, crimes, thefts, lecheries, tortures, the evil deeds of princes, of nations, of private individuals; an orgy of universal atrocity. And it is with this loathsome appetizer that civilized man daily washes down his morning repast.
>
> (Sontag 2003: 96)

Commercial photographer George R. Lawrence gained national attention when he used kites and balloons to lift his specially made camera over earthquake-ravaged San Francisco in 1906 (Holton et al. 2015: 635). As mentioned in the introduction, drone footage had captured the result of the Italian earthquake of 2016 and that year's Japanese earthquakes,[1] a brutal quake that hit Nepal in 2015,[2] bushfires in Australia in 2013, numerous wild fires in America and tornado damage in Arkansas in 2014. When the cruise liner the Costa Concordia ran aground off Italy in 2012, killing 32 people, it was left beached for months. A film production company flew drones over its carcass, with several outlets carrying the footage.[3] When the University of Nebraska's Drones Journalism Lab first put its grant money to use, it was to get footage of its local area, suffering one of the worst droughts in memory.[4] As John Mills, from the University of Central Lancashire's Media Innovation Studio, said: "Aerial footage from natural disasters, conflict zones, public demonstrations or mass sporting events can be a genuinely powerful and unique way to convey a story. It creates a unique dynamic for storytellers. Literally providing a different perspective." But he then warns: "Journalists should remember the power of such footage, and deploy it to maximum editorial advantage rather than let it become 'run of the mill'" (Mancosu 2016). A similar point is made by Sontag in considering the use of television images of suffering, rather than drone footage in particular:

> An image is drained of its force by the way it is used, where and how often it is seen. Images shown on television are by definition images which, sooner or later, one tires. What looks like callousness has its origin in the instability of attention that television is organized to arouse and to satiate by its surfeit of images. Image-glut keeps attention light, mobile, relatively indifferent to content. Image-flow precludes a privileged image. The whole point of television is that one can switch channels, that it is normal to switch channels, to become restless, bored. Consumers droop. They need to be stimulated, jumpstarted, again and again.
>
> (Sontag 2003: 94)

Drone footage can provide that stimulation, as Mills points out above, but it isn't always clear what they add to the narrative of an unfolding disaster. Over Christmas 2015 more than a month's rain fell in a day on parts of Northern England, causing thousands of people to be evacuated as homes were flooded. A serious story, to be sure, and helped up the news agenda that it took place at Christmas when the diary is usually light. There were plenty of individual stories from the sink hole that opened up on a motorway to the city centre streets of Manchester which became rivers. Almost all news outlets had picture galleries of the devastation. Lots chose to feature a 200-year-old pub which straddled the River Irwell but was eventually swept away. The Mirror, a tabloid, headlined its floods story with "Astonishing drone footage of 200-year-old converted pub destroyed by floods." It featured less than a minute of roughly cut clips taken over the pub, which only had three walls remaining. The footage was shot by independent filmmaker David Webster. For the media, one of the imperatives is to be first with the story but here it was the drone footage itself which became the story. The footage, also described as "dramatic", appeared on ITV's website. The gutted building and its precarious position needed no voice over or subtitles to explain, which made it useful for time-pressed editors. The 52 second-long shot showed no change in condition – it was merely observing from a static situation though elevated position. The *Telegraph* also included the footage in its round-up but with the more restrained "Aerial footage shows 200-year-old pub destroyed as River Irwell floods" headline.[5] Two weeks earlier storms had caused Carlisle on the Scottish border to become more like Venice. Drone footage was shot by Juan Latour, who runs a PA hire company in Cumbria but had branched out into drone filming. "Crazy day filming #aerialfootage #CumbriaFloods" he posted on twitter before uploading the footage to Facebook. He then pinged a number of news organizations on his twitter feed who took up his offer to use the footage. The *Telegraph* hosted his video which was headlined as "dramatic" and introduced its story by saying: "Drone footage has emerged which shows the devastating effects of Storm Desmond in Carlisle" (Daunt 2015). It suggests that news organizations now reach for drone footage as a natural part of any disaster coverage, especially now that the number of freelance and amateur operators means there is a steady supply of material.

   The two most recent examples which have done a lot more to showcase the power of drone footage are both linked. One is the huge rise in refugees coming to Europe across the Mediterranean and the other is the war in Syria. We will come to the refugee story later in this chapter, but first Syria. Since 1992 more than 100 journalists have been killed there and in 2015 it was by far the deadliest country in the world for media workers, with 14 killed because of their job.[6] Getting words out of the country let alone images is immensely difficult. The power, then, of footage shot by Russian cameraman Alexander Pushin is undeniable. Material released in 2016 of the battered city of Homs bring home the destruction that has been wrought. The *New York Times* wrote: "The images evoke scenes from a post-apocalyptic video game – an abandoned graveyard, a lone motorcyclist, gutted buildings reduced to pile after pile of rubble – and help explain the desperation driving the country's spasms of civilian flight" (McPhate 2016). It is interesting to note that in trying to describe the footage the writer reaches for a metaphor of a computer game, which is all about a simulation being as real as possible. The power of this footage is that it gives us a privileged view inside a closed system. It is the very lack of simulation which makes it so shocking. Pushin, who has his own drone company, has released footage from other battle fronts in Syria.[7] When the Syrian conflict moved on to Aleppo drones again provided the visual narrative. The footage of the shattered city led the New York Times to liken the images to Berlin in 1945 or Grozny in the 1990s.[8] There is a long tradition of aerial views and warfare; indeed huge technological advances were made as a result of the military deploying cameras above battlefields. After World War One a number of films were made of the blasted landscape to show to the public. *In An Airship Over The Battlefields* by Lucien Le Saint from 1918 reveals the devastation wreaked upon the French and Belgium countryside. Teresa Castro highlights the complex inter-relationship between the images, the viewer and the technology: "a pan-tracking shot that equally emphasises the documentary value of the images of the world it provides and the unique qualities of the alliance between the 'I am seeing' of the camera and the 'I am flying' of the airship." She goes on:

> The fluidity of the camera movement acts undeniably as a source of emotion: emotion linked to the pleasure of discovering the earth from a fresh point of view; emotion attached to the sudden recognition of the land as one more wounded body; and emotion, finally, arising from being able to travel freely through space-time. The cinematographic specificity of this film is fundamental, for no assemblage of aerial photographs could possibly convey, so immediately and effectively, the cenesthetic intensity by the doubled kineticism of the flight and the cinematrographic views.
>
> (Dorrian and Pousin 2013: 125)

A key rationale for the use of drones in warfare is their perceived ability to strike surgically while minimising unwanted casualties. As David Deptula, an American Air Force officer, said: "The real advantage of unmanned aerial systems is that they allow you to project power without projecting vulnerability" (quoted in Chamayou 2015: 12). There is a technological fantasy of being able to solve problems without any side effects. That discourse is infectious. As Sandvik and Lohne write: "We anticipate that a similar rhetoric of surgical precision – and analogous potential pitfalls – will accompany the use of humanitarian drones" (Sandvik and Lohne 2014: 153). There is an increasing wave of claims about the potential for drones to intervene in humanitarian crisis, of which their use by the media is one part. If the answer isn't a drone then you are not asking the right question. There are breathless predictions of drones which can drop medical supplies or give mobile phone coverage to isolated areas. If, in the west, it is about how drones can help us consume ever more thanks to flying pizza deliveries,[9] elsewhere it is about how we can parachute in supplies to benighted countries. One such example was the response to the catastrophic Ebola outbreak in Africa in 2014 where some reached for the drone as a solution.[10] Kristin Bergtora Sandvik, an Associate Professor at the Department of Criminology and Sociology of Law at University of Oslo, wrote of the idea:

> West Africans are strangely absent from the technoscape created by Ebola drone imaginary: it is a technoscape inhabited only by Western actors, who possess hardware, technical skills and the know-how of crisis management. The locals seem to be dead, infected or potentially infected. They are allotted roles as threat subjects (the Ebola terrorist scenario) or victims (in a humanitarian crisis), but either way as individuals and communities mostly void of agency. However, we should remember that while this resonates with the rationales underlying the militarized approach to Ebola, and the determinist views of technology accompanying it; on a different level the militarized approach is also a response to a lack of knowledge about how to deal effectively with disease emerging from structural injustice, a post-conflict context and "culture". Just as drones can't clean up combat, no Ebola Drone can ever "combat" disease.
>
> (Sandvik 2014)

Despite this, drones clearly offer the media not just fresh footage but more importantly safer access. They also extend opportunities to record and interpret events. Matthew Schroyer, founder of the Professional Society of Drone Journalists, said:

> We're starting to see how drones can be quickly deployed to re-create and share important events and places virtually – such as earthquake damage in Nepal and Taiwan, or the Donetsk International Airport in eastern Ukraine. This can be done with existing tech in a matter of hours after a disaster strikes. Outside of journalism, we already see drones being used to quantify the magnitude of man-made and natural disasters. In the future, drones might play an important role in gathering quantifiable data to produce investigative reporting on environmental issues.
>
> (author interview, April 2016)

The Donetsk story Schroyer is talking about saw him create 3D models of Donetsk International Airport using drone footage which helped capture how it had been destroyed in fighting. As Jason Koebler pointed out, footage of destroyed buildings can be somewhat meaningless without context and at the very least gives an idea of what stood there before (Koebler 2015). Much as with Whyld's return to Tacloban, drone footage adds an extra spatial element to a story. The event exists not in isolation but is given some context. Chouliaraki talks of disaster stories as being either adventure news or emergency news and for the former it is a sudden event, briefly captured and then dismissed. A flood, massacre or deadly fire which appears out of nowhere is logged as a dot on a map next to a newsreader's head described in abstract terms and then the news cycle moves on (Chouliaraki 2006). The drone gives an extra temporal context moving the story out of the now and allowing the viewer some measure by which to judge the severity of the event. This is, of course, still a measure set by the limitations and influences of the drone operator but it is an enriching of the story nonetheless. As with Schroyer's work in Donetsk, drone footage is being used to create 3D models of disasters. Within hours of an earthquake hitting Taiwan in February 2016 virtual reality models of damaged buildings had been created using drone footage.[11] The longest running example is probably the International Organization for Migration, which has been using drones since 2012 to capture the damage in Haiti after its 2010 earthquake. Similarly after Nepal's 2015 earthquake, drone footage was used to map damaged historical sites. This was not without its problems, though. As Sharma reported for Reuters:

> With an ability to produce 3-D maps that are often more detailed than satellite imagery, drones can rapidly provide valuable information on the humanitarian needs of survivors. In another example Ben Kreimer team up with BuzzFeed to produce a 360 degree video of the damage caused by a forest fire near San Francisco. The drone carried two different 360-degree cameras to fly over the hazardous area.[12] While supporters of the technology hail its potential to revolutionize humanitarian work, industry experts warn of the security, ethical and legal concerns it raises for agencies and governments. For example in Nepal, media reports of drones flying near security installations stoked fears the footage could be misused.
>
> (Sharma 2016)

The drone is one of those objects which sparks unease across the globe because of its military associations. For people caught up in stressful situations, the appearance and sound of a drone might only raise their anxiety. Another concern is outlined by Goldberg et al.: "It might also be difficult for combatants to identify journalists operating RPAS as civilians and increase the likelihood of being fired upon" (Goldberg et al. 2013: 26).

So in March 2014 a Humanitarian UAV Code of Conduct was drafted to guide the use of drones in humanitarian and development settings.[13] It was put together by UAViators, a network of more than 2,500 people involved in drone use whose mission "is to promote the safe, coordinated and effective use of UAVs for data collection, cargo delivery and communication services in a wide range of humanitarian and development settings".[14] The comprehensive code says: "The use of UAVs to support humanitarian action should be carried out for humanitarian purposes only and with the best interest of affected people and communities in mind, and should adhere to the humanitarian imperative of doing no harm" and covers areas such as data gathering, community engagement and conflict sensitivity. Clearly, journalists are not engaged in humanitarian efforts when reporting from disasters or war zones. However, there is a long tradition of not merely reporting but advocating for help in particular circumstances. The Live Aid effort in the 1980s was sparked by this journalism of attachment from a BBC report from Ethiopia. Journalists are generally conscious of the baggage they bring with them and leave behind when they parachute into such situations. However, a glance at some of the codicils from the Code of Conduct suggests they might be difficult to reconcile with the newsgathering needs of media organizations. For instance the code says drone operators should

> collect and analyse data in a manner that is impartial to avoid discrimination. Informed consent should be secured insofar as the situation allows. As far as possible, data collection and analysis should highlight the needs and aspirations of vulnerable and marginalized groups.

While journalists might notify the authorities they are operating in the area, one finds it unlikely that journalists would seek permission from everyone about the data they collected. Time is of the essence. Tracking down those people is going to be complicated. The footage doesn't really affect them. The simplest solution is to publish and if they complain, take it down. But they are unlikely to complain. Operating in disaster areas is where a journalist's code of ethics, whether culturally internalized or externally imposed by a professional organization, is tested the most because the penalties are the lightest and the oversight the weakest. Therefore the likelihood of journalists behaving other than they would at home is more likely.

Setting aside these ethical issues there are structural limitations to the reporting of crisis abroad which often lead to viewers gaining, at best, only a partial understanding of events and more often reducing stories to abstract tales of woe (Benthall 1993, Chouliaraki 2006, 2013, Minear et al. 1996).

There are advantages to using drones in such situations, as outlined by Astrid Gynnild:

> The main relative advantage for the working conditions of journalists relates to safety issues when working under unpredictable conditions where eyewitness accounts are expected. A dramatic reduction in costs of aerial imagery and videos is, furthermore, a relative advantage in the corporate hunt for sustainable business modules.
>
> (Gynnild 2014: 338)

However it is debatable whether these advantages are enough to overcome built-in cultural or professional preconceptions. An interesting paper by Fair and Parks looked at American TV coverage of the Rwandan refugees in 1996. The footage characterized refugees either in motion or in camps and with little for viewers to connect with. They write: "Relying heavily on close-up and remote images, while neglecting the infinitely more complex historical, and politicized middle-scale, US news organizations, especially television, had no coherent explanation of events happening in and around Rwanda" (Fair and Parks 2001: 37). Aid agencies were desperate to get better coverage that would in turn encourage American intervention in the crisis. They persuaded government officials in Washington to release to the media a selection of images taken by satellites and Navy aircraft. They showed, from a huge distance, dots and blobs which were refugee camps and crowds of people near a vast forest. The media didn't use them. Fair and Parks write:

> News professionals, unlike Refugees International, understood that such images would be unattractive to television because they would need to be decoded and interpreted for viewers. Especially when contrasted to the television camera's ground perspectives, such aerial images reduced refugees to barely visible white and blue dots. . .. the aerial image, rather than providing a kind of gateway into the refugee's soul through a set of wide-open eyes, positions the viewer as a distant and omniscient monitor or displacement and atrocities.
>
> (Fair and Parks 2001: 46)

There was an opportunity for the media to add an extra element to their reporting from Rwanda but it didn't meet the requirements necessary to construct a story. Actually, I am surprised they didn't write the story about aid agencies persuading the American government to declassify military images, which would at least have helped decode the pictures for viewers because it would have been in the context of aerial surveillance. On the one hand, the close-ups "reduce refugees to a part of their bodies – their eyes wide open – in an attempt to symbolize their plight" (Fair and Parks 2001: 41) and on the other the "aerial [images'] data tell little about the embodied circumstances of exile and displacement" (Fair and Parks 2001: 46). Drones may well be able to bridge that spatial gap between the ground images and the satellite images. However, ultimately it still needs to meet the values and construction requirements of media organizations.[15]

The summer of 2015 saw the largest movement of people in Europe since the end of the Second World War. Political instability, climate change, economic collapse and war led to a huge influx of migrants to the continent. It was a complex, long-running story and one where drones were deployed repeatedly by the media. It is worth looking at a couple of examples to see how they were integrated into media coverage. On 5 September the BBC hosted footage of migrants walking along one of Hungary's motorways heading towards the Austrian border.[16] There is no sound other than a music track and some screen text to explain what people are watching. The footage, which lasts one minute four seconds, features several different shots spliced together as the drone pans over the crowd of several hundred people walking along the motorway. Without the on-screen explanation it would be very hard to work out where the people were going to or from or what country they originally left. However, the footage certainly gives a clear picture of the numbers involved in this particular trek. On 23 September there is a more comprehensive report from Matthew Price of the BBC on migrants arriving by boat in Greece.[17] The piece begins with a shot from a drone flying low along a shoreline littered with hundreds of abandoned life preservers. It then cuts to a drone at sea filming as a dingy crammed with people pulls up on a beach. You can hear the sound of people talking and whistling and waves crashing. As the reporter talks about what might lead people to attempt this crossing, the footage cuts to a shoulder-mounted camera on the shoreline right next to people clambering off the boats. There are men, women and children; some are crying, many look exhausted. Price then talks to camera from the now abandoned beach, giving some context, and you realize it is a drone shot as the camera pulls away from him and up to show the amount of debris left behind. The story then moves to the island's port as some of the migrants are buying tickets to get to the Greek mainland. At 2 minutes 30 seconds a migrant gets to say their first word. The reporter introduces a young man from Afghanistan, who isn't named, and says the piece of paper he is holding gives him the right to remain for a month. The reporter says the man is heading for the mainland and then asks him to confirm that after that he intends to go to Germany. He gets the migrant to repeat the Afghan word for Germany. In a rapid series of cuts, other migrants, a mixture of men and women, say their country of origin to the camera. A similar set of cuts then has them saying the country they are heading for. Finally the drone is used to film above several hundred migrants waiting to board a ferry as the report closes after 3 minutes and 42 seconds. In October, footage shot by AP photographer Darko Bandic is carried by several outlets, including the *Telegraph*.[18] It shows hundreds of migrants walking along a country road in Slovenia close to the border with Croatia. The 53 second film has no sound but screen text explaining what people are seeing and that Slovenia has complained about a lack of man-power to deal with the migrants. The footage is a mix of different shots; some very high showing the people strung along over a mile of road, others about 20 feet off the ground and running alongside the column. No one acknowledges the drone and it is hard to make out individuals. In all three cases the fact that a drone has been used is flagged up in the headline or sub headline. It should be noted that Bandic, as with many agency photographers, spent much of 2015 travelling around Europe recording the stories of migrants. They would have been well informed and committed to the story and certainly not simply parachuted in to get a simple image.

In *The Spectatorship of Suffering*, Chouliaraki analyses a news segment from 2002 of migrants being rescued from a boat in the Mediterranean. The piece is made up of footage from a camera on a Maltese lifeboat to provide a dramatic eyewitness account. Chouliaraki classifies this type of report as emergency news – one which demands action:

> This appeal to action may well work as a symbolic vehicle that gestures towards token engagement only with distant sufferers and, ultimately, feeds on the spectator's denial and fatigue. Instrumental in a regime of pity that humanizes sufferers and encourages reflexive spectators is the quest *to historize the scene of the suffering*.
>
> (Chouliaraki 2006: 150)

The story of the migrants in 2015 is one where the drone footage undoubtedly turns the story into something more spectacular. We get a much clearer idea of the scale of the specific incident. We are given a privileged eyewitness perspective; truly god-like as we look down upon the people walking beneath us. The god-like view does not, though, give us omniscience. We don't gain any extra knowledge about who the people are. They do not even seem aware of the camera above them and there is certainly no way they can engage with it. There is no chance for them to influence what is being shot – they are passive participants in the drama.

At times in these reports we are unaware of the camera, as is usual with news packages. At other times, especially as it pulls away and up, we are treated to something more like a cinematic experience. Our relationship with the piece changes; indeed they have been signalled as drone footage so there is an expectation for the spectacular. At a high altitude with a slow flight path it mimics the view out of an aeroplane window. If the camera moves off the horizontal plan we become more conscious of the mechanism that allows us this viewpoint. A swoop down signals to the viewer that this is not a helicopter shot but something we would have otherwise have seen in a film. One could argue that our conditioned reaction then gets confused. An adventure movie excites and thrills but we are not generally engaged by them. We consume and then we move on. That movie reaction is particularly so with the reveal; the slow rise by the drone and then the move across a landscape. Here we are made complicit in the spectacle. The gasp-inducing rise above a devastated city excites that reaction because of its filmic nature allied with the fact that what we are looking at is 'true'. There are no blue screens or models; those blasted buildings exist in a place we are unlikely to visit. We are put in the position of a voyeur: "a witness who has been freed from the moral obligation to act and so can sit back and enjoy the high-adrenaline spectacle unfolding on the screen" (Chouliaraki 2006: 145).

The optimistic reading would be that drone footage offers a greater range of options for editors to tell stories. Where once a foreign disaster might warrant only brief coverage because of the paucity of images, now it might gain a more prominent slot if there is drone footage. However, news is always made up of light and shade. Items of varying length and tone (within certain boundaries). A series of three-minute, hard-hitting reports would become indigestible after a while – the news feast requires some variety in the dishes being served. Indeed one might argue that small wars, far away, may struggle even harder to gain attention if drone footage privileges suffering elsewhere.[19] As Chouliaraki writes: "Despite the expansion in news delivery technologies, all pieces of news are eventually subject to a process of selection and symbolic particularization that defines whose suffering matters to Western spectators" (Chouliaraki 2006: 187).

Notes

1 Footage can be found here: https://www.washingtonpost.com/news/capital-weather-gang/wp/2016/04/18/drone-video-over-japan-earthquake-site-shows-massive-landslides-fractured-earth/
2 See this BBC site for gallery of footage: http://www.bbc.co.uk/news/world-asia-32575991
3 See Time magazine's website for instance: http://newsfeed.time.com/2013/07/17/watch-dramatic-drone-footage-of-costa-concordia/
4 Footage can be found here: https://www.youtube.com/watch?v=HV0iKlF9AdA
5 Footage can be found here: http://www.telegraph.co.uk/news/weather/12069952/200-year-old-pub-collapses-as-River-Irwell-floods.html and http://www.mirror.co.uk/news/uk-news/astonishing-drone-footage-200-year-7076521

6 Figures from the Committee to Protect Journalists. See https://cpj.org/killed/mideast/syria/

7 For details see http://www.nytimes.com/2015/10/21/world/middleeast/watching-syrias-war-as-a-music-video-shot-by-a-russian-drone.html and for selection of videos see the Drone Works YouTube channel: https://www.youtube.com/channel/UC8AClJusxRD3vuxcPCdYaEA

8 see: http://www.nytimes.com/2016/10/15/world/middleeast/aleppo-destruction-drone-video.html?smid=tw-nytimes&smtyp=cur&_r=0 and for images from a picture agency in the city see: http://aa.com.tr/en/pg/photo-gallery/anadolu-agency-photos-of-aleppo-evac-on-international-press/146/291568

9 Though Amazon is the best-known there are many companies reportedly developing drone deliveries. What gets less attention are the logistical and legal problems. For the latter see: difficult http://jrupprechtlaw.com/amazon-drone-delivery-3-major-legal-problems-amazon-prime-air

10 See for instance http://www.bbc.co.uk/news/technology-29945401

11 See https://irevolutions.org/2016/02/07/aerial-robotics-virtual-reality/ for details

12 You can see the footage here: https://www.facebook.com/BuzzFeedVideo/videos/1869492456524961/

13 You can see a copy of it in full here: https://docs.google.com/document/d/1Uez75_qmIVMxY35OzqMd_HPzSf-Ey43lJ_mye-kEEpQ/mobilebasic?pli=1

14 You can find out more about UAViators here: http://uaviators.org/

15 In an updated effort similar to the use of satellite images, in November 2016 the UNHCR used drones to record the position of displaced people in Africa and shared with the medis. See: http://www.unhcr.org/news/latest/2016/11/582dc6d24/unhcr-uses-drones-help-displaced-populations-africa.html

16 See http://www.bbc.co.uk/news/world-europe-34165674

17 See http://www.bbc.co.uk/news/world-europe-34333215

18 See http://www.telegraph.co.uk/news/worldnews/europe/slovenia/11954678/Migrant-crisis-Drone-footage-shows-the-flow-of-migrants-in-Slovenia.html

19 Or as the Taliban showed at the end of 2016 when it used a drone to film one of its own suicide bombings and released the footage online, it offers a new way to grab attention. See: http://www.aljazeera.com/news/2016/10/taliban-drones-film-attacks-afghanistan-161023061347421.html

References

Benthall, J. (1993) *Disasters, relief and the Media.* London: IB Tauris.

Chamayou, G. (2015) *Drone theory.* London: Penguin.

Chouliaraki, L. (2006) *The spectatorship of suffering.* London: Sage.

Chouliaraki, L. (2013) *The ironic spectator: Solidarity in the age of post-humanitarism.* London: Polity.

Daunt, J. (2015) Drone footage shows floods in Carlisle. *Daily Telegraph.* 6 December 2015. http://www.telegraph.co.uk/news/weather/12036038/Drone-footage-shows-flooding-in-Carlisle.html

Dorrian, M. and Pousin, F., ed. (2013) *Seeing from above: The aerial view in visual culture.* London: IB Taurus.

Dupuy, L. (2013) Roundup: Drone journalism in Tacloban. *Emergency Journalism.* 9 December 2013. http://emergencyjournalism.net/roundup-drone-journalism-in-tacloban/

Fair, J. and Parks, L. (2001) Africa on camera: Television news coverage and aerial imaging of Rwandan refugees. *Africa Today* Vol. 48(2), 35–57.

Goldberg, D.; Corcoran, M. and Picard, R. G. (2013) Remotely piloted aircraft systems & journalism opportunities and challenges of drones in news gathering. *Reuters Institute for the Study of Journalism.* https://reutersinstitute.politics.ox.ac.uk/fileadmin/documents/Publications/Working_Papers/Remotely_Piloted_Aircraft_and_Journalism.pdf

Kaufman, L. and Somaiya, R. (2013) Drones offer journalists a wider view. *New York Times.* 24 November 2013. http://www.nytimes.com/2013/11/25/business/media/drones-offer-journalists-a-wider-view.html?_r=1&hp=&adxnnl=1&pagewanted=1&adxnnlx=1385479630-qaDKCPHyTwlgrDdZG8U5Lw

Koebler, J. (2015) 3D maps made with drone footage show the destruction of a Ukrainian airport. *Motherboard.* 20 January 2015. http://motherboard.vice.com/read/3d-maps-made-with-drone-footage-show-the-destruction-of-a-ukrainian-airport

Mancosu, M. (2016) Opportunities and challenges in drone journalism: 15 industry experts share their views. *Skytango.* 29 April 2016. https://skytango.com/drone-journalism-opportunities-and-challenges-15-experts-share-their-views/#top_5_opportunities_for_drones_in_journalism

McPhate, M. (2016) Drone captures view of a devastated Syrian city. *New York Times.* 3 February 2016. http://www.nytimes.com/2016/02/04/world/middleeast/drone-captures-view-of-devastation-in-homs-syria.html?_r=0

Minear, L.; Scott, C. and Weiss, T. (1996) *The news media, civil war and humanitarian action* Boulder: Lynne Reinner.

Penhault, K. (2013) CNN uses drone cameras for aerial view of Tacloban. *CNN Press Room.* 18 November 2013. http://cnnpressroom.blogs.cnn.com/2013/11/18/cnn-uses-drone-camera-for-aerial-view-of-tacloban/

Sandvik, K. (2014) Fighting the war with the Ebola drone. *Norwegian Centre for Humanitarian Studies.* 3 December 2014. http://www.humanitarianstudies.no/tag/drone/

Sandvik, K. and Lohne, K. (2014) The rise of the humanitarian drone: Giving content to an emerging concept. *Millennium: Journal of International Studies* Vol. 43(1), 145–164.

Sharma, G. (2016) Armed with drones, aid workers seek faster response to earthquakes, floods. *Thompson Reuters Foundation.* 15 May 2016. http://www.reuters.com/article/us-humanitarian-summit-nepal-drones-idUSKCN0Y7003

Sontag, S. (2003) *Regarding the pain of others.* London: Penguin.

Whyld, L. and Phillips, T. (2014) Typhoon Haiyan: Aerial footage shows how Tacloban has recovered six months on from the natural disaster. *Daily Telegraph.* 6 May 2014. http://www.telegraph.co.uk/news/worldnews/asia/philippines/10804117/Typhoon-Haiyan-aerial-footage-shows-how-Tacloban-has-recovered-six-months-on-from-the-natural-disaster.html

5 Investigations and activism

When it came to investigating issues within the agriculture industry in America, journalist Will Potter had one serious legal obstacle. In some states a so-called Ag-Gag law has been enacted which specifically forbids the use of undercover material gathered on farms. It has long been necessary to expose wrongdoing in such closed environments where official documents tell little to go in secretly (Schiffrin 2014).[1] Despite what seems a clear violation of first amendment free speech rights, the Ag-Gag law presented a huge hurdle for journalists. So Potter decided that rather than going in undercover, he would go over the fence. Using an off-the-shelf drone but with his own modifications to the camera, Potter shot footage in Washington, Oregon and Michigan. "I am a print journalist by training," he said. "Coming to this new technology, it was a new world. But I grew up as a video gamer. How difficult would it be? You just have to control the camera, you have to control the drone and try and get good quality pictures" (author interview, February 2016). Of course, as Potter discovered, it was more complicated than that: "You have to know the law and your rights. But just because you explain that doesn't mean the police are necessarily going to treat you with respect. So you have to be cautious and prepared." Investigative journalists often find themselves working closer with or against lawyers than many of their colleagues on other beats. The stories tend to make allegations of improper behaviour against powerful individuals or organizations who will readily defend themselves in court if necessary. The techniques used to construct such stories are more likely to use intrusive methods to gain evidence. The public-interest test necessary to satisfy regulators that an invasion of privacy or subterfuge was required is about exposing wrongdoing and no other options were available. Most stories are not going to meet that criteria. As Hunter described it:

> Investigative journalism involves exposing to the public matters that
> are concealed – either deliberately by someone in a position of power,

or accidentally, behind a chaotic mass of facts and circumstances that obscure understanding. It requires using both secret and open sources and documents. Conventional news reporting depends largely and sometimes entirely on materials provided by others (such as police, governments, companies, etc.); it is fundamentally reactive, if not passive. Investigative reporting, in contrast, depends on material gathered or generated through the reporter's own initiative (which is why it is often called "enterprise reporting").

(Hunter 2011: 10)

As we have touched upon, and will do so in more detail later, the laws surrounding drones are complex. Also, as it is a new technology, many are unclear about exactly where the boundaries lie. I would argue that, as investigative journalists spend a lot of time making themselves familiar with legal constraints and negotiating with the police and regulators about what they can do, operating drones may present technical challenges but the legal problems are just another day at the office.[2] As Potter describes it: "You don't want to loiter around because of hostility from farmers or the police. I got followed quite a bit."[3] His advice is to carry out such work in a group:

You are in a highly pressured situation and you have to focus on the drone. Getting quality video and controlling it and being safe makes you vulnerable. There is also the safety of anyone if you have an accident. As a journalist, having multiple people with you is very useful.

(author interview, February 2016)

In the end Potter says that he got some very good footage showing the state of farms, including pools of faeces that stretched for acres, and it very much added to his story. However he added: "I do not think it is universally applicable. There are very serious conversations about privacy and regulation. But there is quite a lot of potential." He added:

They're increasingly affordable, safe and easy to launch and the possibilities are really promising. . . .. I just think we have to be responsible with how we use this technology – to try to use it to the best of our ability and add to the story, rather than just use it because it's a fun new tool.

(author interview, February 2016)

Potter was not the first person to use drones to specifically look at animal welfare. In Australia the Animal Liberation group announced in 2013 that it would be flying one over free-range egg farms, sheep farms and cattle yards to gather evidence of abuse. Mark Pearson, an executive director for the group, told *The Age*:

> If an egg producer says that they are free-range, it would be helpful to check their claims by filming from above the property. You can gather the evidence, and there's no need to trespass. Or, let's say we find a sheep dying from fly strike, we can record the location on a GPS and notify the authorities.
>
> (Cubby 2013)

Farmers were, unsurprisingly, less enthusiastic, citing it as an invasion of privacy, but Pearson said: "People are entitled to know and see what's going on." The activists had reportedly spent $14,000 buying a drone and sent operators on a training course before they could fly it. In this chapter we'll consider the use of drones for investigative journalism and by activists, and it is deliberate that these are covered together. Some of the first recorded instances of drones being used to gather information to inform the public involve either activists or investigative journalists and the former have done much to test the technical and legal boundaries of what is possible. One of the first examples didn't involve a journalist at all. In 2011 a man from Texas was flying his new drone near Dallas when he noticed that a river by an old meat packing plant was heavily discoloured. "I was looking at images after the flight that showed a blood red creek and was thinking, could this really be what I think it is? Can you really do that, surely not?" (quoted in Keneally 2012). He alerted the coastguard and several agencies launched an investigation. It turned out the plant was leaking pig blood into the creek, which turned red a major river. Meanwhile various studies have looked at the different possible journalistic applications of drones and aside from covering disasters (as discussed in the last chapter) their investigative potential is the other aspect often highlighted. Goldberg et al. warn their use could move journalists closer to that of private detectives:

> Aerial platforms, for example, would make it possible to discretely follow public officials or others to clandestine meetings or to hover outside windows photographing or even listening in on meetings. With high definition cameras, it might be possible even to photograph documents at a distance.
>
> (Goldberg et al. 2013: 24)

Meanwhile in their analysis of eight early instances of drone journalism, Tremayne and Clark found that their use by citizen reporters and activists was crucial. They write:

> These instances of drone usage demonstrate . . . [that] once again technology is helping shape the coverage of news. In four of the eight cases mentioned above, non-mainstream journalists, or citizen journalists, were responsible for recording video. Several of these cases were activists trying to enhance media coverage of their causes.
>
> (Tremayne and Clark 2014: 238)

The authors conclude: "drone journalism may be another example of new media technology being used by citizens to advance their social and political causes and, purposefully or not, contesting the power of traditional media" (Tremayne and Clark 2014: 239). While not suggesting the investigative journalist and activist or citizen reporters are the same, both have been early adopters of this technology and their experiments suggest how drones might be better used for newsgathering. They also share some methodological approaches in terms of targeting issues and gathering evidence. As Hunter writes:

> Conventional news reporting aims to create an objective image of the world as it is. Investigative reporting uses objectively true material – that is, facts that any reasonable observer would agree are true – toward the subjective goal of reforming the world. That is not a license to lie in a good cause. It is a responsibility, to learn the truth so that the world can change.
>
> (Hunter 2011: 8)

And increasingly, pressure groups, NGOs and other third sector organizations are developing investigative techniques and selling the results on to the media to help provide news coverage.

Since the business model for the news media was turned upside down with the rise in free online information, one of the victims is seen to be investigative journalism. If print newspapers in America and the United Kingdom were axing staff and moving from daily to weekly or simply folding altogether, then what hope was there for in-depth reporting? If the commercial television market and public service broadcasting was under attack from streamed services and changing ideological climate while advertisers found other ways to reach eyeballs, what future was there in investigative programmes? Investigative journalism is expensive and time consuming.

An investment in researching a story does not always lead to an output which can be published. Since investigative journalism tended, though by no means exclusively, to see its targets as government and corporate malfeasance, the assumption was that there were few tears shed in parliaments and boardrooms if fewer such investigations took place. However, a report from the House of Lords select committee on communications into the future of investigative journalism said: "We heard much evidence which painted a pessimistic picture of the economic problems facing investigative journalism but we have heard no evidence that leads us to conclude that investigative journalism will disappear: we believe that it will continue" (House of Lords 2012: 72). The committee heard many different financial solutions suggested that might support investigative journalism but the belief etched in that conclusion was more faith based than fact based. There has certainly been a rise in philanthropic donations and foundations willing to support efforts. The decline in institutional investigative teams has been mirrored (if not matched) by a rise in independent outfits. In the United Kingdom, The Bureau of Investigative Journalism is one such example and has succeeded in placing noteworthy stories in numerous outlets.[4] Brant Houston, in his analysis on the future of investigative journalism, said:

> Investigative journalists, because of the dedication and the zealotry they bring to their work, persist in the worst of times; sometimes they seemingly flourish when the challenges are greatest. "The people who are drawn to it and perform it are so dogged they are not going away," said Tom Casciato, the executive producer of the PBS series Expose, which has profiled investigative reports and the journalists doing them since 2006. "They got into it because they think it's important. They can't not do it." The result is that reporters and editors themselves have provided the models for how investigative journalism can proceed into the future.
>
> (Houston 2010: 47)

There is an element of finger-crossing here, hoping that a ready supply of zealots will always provide stories whatever the cost. One also wonders about what type of stories such zealots wish to pursue. McChesney mocks the notion of a faith in the market to provide:

> Journalism is the main way modern societies produce and disseminate political information, and it is of singular importance in democracies. Much has been made of how the Internet has destroyed the commercial media business model. With no sense of irony, the same people argue

that the Internet will combine with free markets to magically re-create a new, different and superior new media system sometime in the future.

(McChesney 2013: 21)

Investigative journalism has fractured so that rather than being simply the preserve of large mainstream corporations, there are freelancers, start-ups, activist organizations and academics all picking up the baton and supplying the demand – and they've been some of the first to adopt drones.

One of the first examples of a drone being used by journalists was actually for investigative purposes. It was deployed by Channel 9 reporters in Australia in 2011. The government there was holding refugees trying to enter the country in various detention centres including Christmas Island. Concerns had been raised about the conditions at the camp and so reporters from Channel 9's *60 Minutes* programme went to ask questions. As Goldberg et al. describe:

> After being denied entry to the facility, the *60 Minutes* team launched a multi-rotor to record exclusive, if somewhat unremarkable, aerial images of the detention centre, before the craft crashed into the sea. Following complaints from the Immigration Department, the Australian Federal Police investigated the incident but no laws appeared to have been broken and the Australian civil aviation authorities declined to pursue the matter.
>
> (Goldberg et al. 2013: 22)

Reporter Liam Bartlett said:

> We'd tried the front door without success, so this was the only way to show how and where asylum seekers were detained – a bird's eye view from an unmanned camera. It's unconventional but I think it's the only chance we've got of being able to see inside.
>
> (quoted in Tremayne and Clark 2014: 237)

In terms of meeting the requirements for intrusion, the Christmas Island case appears to tick all the boxes. The reporters had been denied permission, there was no other way to look at the area, they were following up on reported complaints rather than going on a 'fishing' exercise and it was a matter of public concern – both the conditions in the camp and the government's refusal to allow access. Sometimes it doesn't need to be so sneaky. Television producer Joe Myserclough recalls that his programme, which featured drones flying over the United Kingdom to show previously unseen sights, nearly landed an unlikely scoop:

> We really wanted to fly around the radomes [giant golf balls] at Menwith Hill spy base in Yorkshire. The base is jointly run by the RAF and the US intelligence service, the NSA. At one point, the RAF said yes, but then the NSA said no. Then the NSA changed their mind and said yes, but the RAF also changed their mind and said no. One day they might both say yes on the same day and we'll have to get a drone in the air before they change their minds.
>
> (author interview, July 2016)

The first time the *New York Times* put a drone it owned to use, as opposed to using footage shot by others, was to record the eerie beauty of the Greenland ice pack. The story was about the melting of the ice sheet and one might question whether it meets the definition of investigative journalism. However, it certainly helped bring to attention an issue of public concern and it is not a story that is on the regular news diary nor is it a sudden event like an earthquake.[5] It required time and skill to tell and the objective was to get people to act differently. Staff photographer Josh Haner writes about the process of carrying out this assignment, which suggests some of the same lessons from Will Potter's work in America. For one thing, the limited flight time means that a lot of planning is required to ensure flights can be as efficient as possible. For another, working in at least pairs where the terrain is difficult (that might be legally for Will or environmentally for Haner) is recommended. The concentration required to fly the drone and ensure it is getting the right footage makes people oblivious to what might be going on around them. Both Potter and Haner suffered crashes as they rushed to land their drones because they needed to leave quickly. For Haner this was a chance to draw readers into the story as well as showing clearly the reality of the melting glaciers:

> I wanted to take the reader from the meltwater lake and past the researchers as they pulled their floating sensors across the river, then trace the river almost a mile to where it disappears into a moulin, a giant hole in the ice that drains the water through tunnels in the ice sheet out into the ocean.
>
> (Haner 2015)

As Matthew Schroyer said: "In the future, drones might play an important role in gathering quantifiable data to produce investigative reporting on environmental issues" (author interview, April 2016). The *New York Times* article is one example. Another is the work of Dickens Olewe and Ben Kreimer documenting the Dandora Dump in Nairobi. The pair flew a drone over the site and then created a 3D model so that people could fully appreciate its 30-acre scale and how close it sat next to schools and houses. Olewe said: "I thought it was really important to inspire the imagination of not just our readers, but also the government, who I knew at some point would want to regulate this space" (Edge 2015). In the United Kingdom, the BBC's premier investigative television programme 'Panorama' has repeatedly used drones to add interest to its stories. Generally they have provided supplementary footage rather than show anything new, though in a story about flooding in 2014 a drone shot as the sun rose over the flooded Somerset levels arguably conveyed a lot quicker and to stunning effect the impact of the rains. Pilot Dave Halton from Lancashire-based AerialVue was commissioned to do the job and said the inclement weather and very short production time made for a hectic but ultimately successful shoot. "With a full moon it was a very eerie and quite emotional sight," he said.[6]

Sometimes, though, it appears that members of the public are quicker and more imaginative than journalists. As with the example from Texas of the river pollution, it is the information provided by interested citizens that is setting the news agenda. However, unlike for instance mobile phone footage from an accident where the eyewitness is there by chance and happens to have the facility to record, some people are actively thinking about applications for drones and how they might be used to influence an issue. In Atlanta residents in a neighbourhood which has suffered because the economic downturn has led to properties being abandoned and fallen into disrepair have taken to the air to illustrate the problem. Alan Holmes, who lives in the Oakland City neighbourhood, said that people had become immune to the problem because they saw it so regularly. The idea was that the drone footage would offer a new perspective and get people thinking differently.[7] The shots over the dilapidated houses offer an eerie view of the desolation and mimic the famous portraits from Detroit which recorded historic buildings in decline. "People are never able to see the footage from that angle, from that level. I think a lot of people who walk past the abandoned houses and have lived in the neighborhood 15, 20 years almost get used to seeing it," said Holmes (quoted in Alexander 2016). The footage was picked up by a local TV station and that prompted a statement from the mayor. This was a story which could have been done at any time by journalists, assuming the right paperwork had been completed, but it took people motivated by a particular concern to make it happen.

The year 2011 was the year activists began deploying drones. In that year a huge demonstration took place in Warsaw and as police held their lines in the capital a drone took off from behind them. Soaring hundreds of feet into the air it cruised over a central square where demonstrators were corralled. Police reinforcements can be seen moving up and flares being thrown. A second set of footage from the same demonstration sees the drone take off over the heads of police officers heading towards the demonstration but they don't give it a glance.[8] It is a remarkable example of activist footage and one that was repeated in America that same year. As Culver reports:

> Tim Pool, an Occupy Wall Street Protestor, flew what he dubbed the "Occupator" over police-restricted areas to document events when he could not gain physical access for video recording in person. He was able to navigate the device into "frozen zones" – where the police locked out the news media and anyone with cameras – and record police action against protestors.
>
> (Culver 2014: 54)

In Russia, Air Pano launched a hexacopter to show a crowd in Moscow in 2011. A year later one was flown over an anti-government rally in Argentina (Goldberg et al. 2013: 21). In 2013 the anti-government demonstrations in Turkey were recorded by a drone.[9] No journalist in the United Kingdom or America would likely get permission for such coverage because of the safety concerns for those below should the drone malfunction in flight. That hasn't stopped demonstrations regularly being filmed from above in various cities by both activists and freelance journalists – sometimes with them wearing the same hats.

During protests in Bahrain in 2006, activists used Google Earth to source images of the homes of elite members of the government to highlight their luxury and poverty of their subjects. In response, the government blocked Google Earth but protestors circulated PDFs of the images to get round the ban. As one activist said: "Some palaces take up more space than three or four villages nearby and block access to the sea for fishermen. People knew this already. But they never saw it. All they saw were the surrounding walls" (quoted in Gutermuth). It is a good example of how the observation from above can provide not just a new perspective but also new information. In this case the means to control the flow of images still favoured the government. They could shut down the access Bahrain-based computers had to the internet. It was a technique which regimes deployed frequently during the uprisings across North Africa and the Middle East. A decade later and half a world away, another community closed off to the outside was opened up from above. The gated compound at Holercani in Moldova covers more than 160 acres and is guarded by the military. While the former Russian republic suffers from grinding poverty, some do a lot better than others and some of these live in Holercani's gilded retreat. A glimpse of their lifestyle was offered in February 2016 when journalists flew a drone over its walls to reveal the tennis courts and mansions. The footage was shot by a team from a news organization called Agora[10] and they had learnt their skills at a training workshop run by colleagues from the Ukraine and Romania who had already cut their teeth covering demonstrations and other events. The session was put on by the Independent Journalism Centre supported by Internews. Those taking part could then put in story ideas and the best won a drone thanks to Radio Free Europe. Constantin Celac was one of those who shot the footage. He said:

> As a child in Moldova, every boy dreams of being a pilot. Growing up, my dreams and desires have changed. Now I dream of using the power of journalism to send a message to the people who govern us, to make a difference. Producing such a report with a drone, well, it's a perfect realization of my childhood dream and my adult reality.
>
> (quoted in Rouse 2016)

The journalists were helped by the fact that, at the time Moldova did not have any laws regulating drone use, it had classed information about Holercani a state secret. The episode is an interesting example of where outside organizations are providing training and basic kit for journalists from less-developed nations. The operation of a drone does not require a huge amount of infrastructure or technical support, which means that it is a technology which can be sustained in relatively poor countries. In some ways the ability to investigate using drones is even greater because, as in Moldova, there aren't the legal restrictions in the first place. Of course, sanctions for journalists who displease the authorities in such countries can also be a lot more severe and arbitrary. As Gynnild writes:

> A main factor in this emergent process towards explorative journalistic work practices is the increasingly transparent and trans-disciplinary mode of sharing and collective experimenting on the internet. The global accessibility to information about drones as a journalistic innovation provides new opportunities for early adopters, both within and outside of journalism, to connect, network, and instantly learn from each others' experiments and experiences.
>
> (Gynnild 2014: 341)

That sharing economy is supplemented by regular training sessions provided by (often Western) philanthropic journalistic organizations so people can get their hands on the technology and experiment.[11]

The blurring of lines between journalist, activist, professional and amateur is exemplified by the use of drones which have been adopted by people wearing all of those labels – sometimes at the same time. In the spring of 2016, the campaign group Greenpeace put out an advert for an investigative journalist to join its EnergyDesk team. "We're building a team of talented investigators tasked with finding the great, untold environmental stories," it said. "You'll be a key part of a project to embed journalism and investigations in Greenpeace's global work – working with colleagues in China, India, Indonesia, the US and elsewhere and using a range of tools from drones and satellite images to our new Bloomberg terminal." Whoever got the job was also promised: "As well as partnering with established media organisations you will be responsible for publishing your own work on our own editorially independent platform, EnergyDesk." EnergyDesk has broken many stories from palm oil investigations to farm subsidies and nuclear power safety concerns.[12] It appears that images used to illustrate several stories have been taken from drones, though they do not seem to be a fundamental part of the newsgathering process. In the civil sector there is huge growth in the market for drones as observation tools for monitoring construction sites, oil rigs, ships, planes and any other large and expensive item which might otherwise be difficult to access. Such aerial platforms can also be turned into more critical eyes with the information gathered assessed for problems to be highlighted rather than solved. The Greenpeace advert helps signal the evolution of investigative reporting mentioned previously and the explicit reference to partnering with established media organizations suggests that the NGO is more than a source but actively sets the news agenda. It also highlights one area of real drone innovation in the civil sector which has implications for journalism, and that is its use in environmental reporting.

The examples from Greenland and Nairobi mentioned above show how journalists have used drones to investigative environmental issues but conservation groups have shown there is far greater potential. Schroyer said:

> In Kyrgyzstan in 2015, the Swiss Foundation for Mine Action used a fixed-wing drone to find the volume of toxic mine tailings which was valuable data for a remediation project. Additionally, Precision Hawk, a company that specializes in data collection using fixed-wing mapping drones, collaborated with Roboticists Without Borders to measure the volume of a toxic waste site that was actively burning in South Carolina. These same approaches could be used by journalists to assess the severity or cost of environmental disasters when there is no official

effort to do so, to fact-check official reports, or to gain information that has not been made public.

(author interview, April 2016)

Prof Serge Wich is a primate biologist from Liverpool John Moores University and also one of the founders of Conservation Drones.[13] It looks to work principally in developing countries to develop the use of low-cost drones for conservation-related applications. The initiative started when Prof Wich met conservation ecologist Lian Pin Koh in 2011 to discuss the issue of wildlife conservation in Southeast Asia. The idea was to use drones, but at that stage commercial drones were too expensive, so the pair built their own. In 2012 they took it to Indonesia and carried out field tests, collecting images as it flew more than 30 missions over the forests. The project has taken off from there. A key issue has always been to make the technology available at a local level. Rather than import technology which might be expensive to maintain, the aim has been to build up a base of expertise in the field so that it is sustaining in the long term. According to Prof Wich, now that the practicalities of flight have been demonstrated the potential uses are expanding. "The focus has been on the visual image but now we are seeing a shift. Now we are looking at multi-spectral sensors. Over the past few years they have become a lot lighter and more affordable" (author interview, February 2016). Wich sees a huge amount of potential for environmental reporting. "The key point is they are very flexible. Currently, some struggle with the regulations but generally we are seeing a lot of movement to make use of these" (author interview, February 2016). The most common use by the media when it comes to the environment is by wildlife filmmakers who have readily adopted drones to get footage. The difficulty of getting up close to animals has meant that natural history programmes have always been imaginative when it comes to using cameras. They are often disguised and operated remotely so that the target animals can be recorded without disturbance. Drones are not, then, much of a leap for these practitioners in terms of the technology or applications. Charlotte Scott, from the BBC's Natural History Unit, said: "In the old days, we had cameramen hanging out of the side of a helicopter. But now, you can be on a boat in the ocean and you can send a little octocopter up and film dolphins swimming in the ocean." Scott recalls filming in China in 2008 trying to record a pod of dolphins. There were restrictions on how close they could get which would now be solved by using a drone. "It really opens up what you can film," she said (quoted in Youngs 2014).

One example of the application for conservation given by Prof Wich involved a drone being flown alongside a river in Indonesia. Loggers had illegally cleared forest parallel to the river but left a deep strip intact so that from a boat it looked as if nothing had changed. An overflight by a drone revealed the truth. "It was a real eye opening and it was not something the loggers were expecting," said Prof Wich (quoted in Youngs 2014). It is not an isolated example. Efforts by conservationists around the globe have shown several distinct advantages to using drones. In Guyana, people used tutorials off YouTube to construct a drone which could demonstrate illegal logging in their area (Wallis 2016). In Mexico scientists heard that illegal logging had taken place in a biosphere set up to protect a particular species of butterfly. They were forbidden from entering the area, so employed a filmmaker to fly a drone over the site. This revealed that ten hectares had been devastated, causing huge damage to the vulnerable monarch butterfly population (Platt 2016). A paper by Gonzalez et al. records projects using drones for "monitoring sea turtles, black bears, large land mammals (e.g. elephants), marine mammals (e.g. dugongs), birds (e.g. flocks of snow geese), wildlife radio collar tracking and supporting anti-poaching" (Gonzalez et al. 2016: 2). We will consider the anti-poaching use shortly but the paper tries to address a specific issue around the collection of data. Simply put, the volume of material recorded and the time taken to process means that sometimes it is no quicker than traditional surveying methods. A lot of work is going to develop algorithms that can quickly identify targets in the field, so speeding up the process can predict where creatures might be so that the short flight times are used as efficiently as possible. Increasingly the use of automative predictive software is seen as reducing the brute force needed to carry out any analysis. The technique tested by Gonzalez and his team in Australia flew a drone on automatic over a forest with a normal video camera and a thermal camera and then used detection algorithms to try and identify from the footage how many koalas were in the area from the huge amount of data gathered and with other animals confusing the returns. Having previously counted on the ground, the team knew what score they should record – and the drone was 100% accurate. One might see where such predictive technology might also be developed for military purposes. Indeed journalists using drones to report on crowds of demonstrators could use software designed for monitoring herds to similar effect in terms of assessing the numbers and make-up of participants.

Such predictive algorithms have also been used as part of anti-poaching efforts. Drones have been used to protect animals in Zimbabwe, Zambia, Kenya, Tanzania, Namibia and South Africa. In the latter country a project linking AirShephard, a nonprofit focussed on aerial solutions to poaching, and the University of Maryland cut rhino poaching entirely in one area that previously had lost as many as 19 rhinos a month (Raxter and Young 2015). A project involving the WWF in Namibia had reportedly achieved a 95% success rate in predicting the location of poaches and target animals (Andrews 2014).[14] Much like burglars will look for certain signs that indicate a potential house to rob, poachers had patterns of behaviour that could be programmed into drone flight plans. While it sounds like a simple solution, concerns have been raised that are worth summarizing because that same sense of technological awe can colour the discourse of drones in journalism.[15] Setting aside the noise of drones which might upset animals, the specific drones used in anti-poaching can be very expensive. The infrastructure needed to keep them going and analyse their returns means providing effective coverage for all Africa's national parks would run into the millions. As Conservation Drones has aimed to do, there is a sense that technological solutions are effective when developed and embedded from the ground up rather than imposed from outside. The control and sustainability of such programmes needs to last after the academics from universities in the west have moved on to other projects. Google has donated some $5 million dollars towards the anti-poaching efforts in Namibia for reasons which are surely entirely philanthropic. However images of exotic animals at risk are very powerful and there is an easy narrative of good guys and bad guys with the smart technologists helping the willing amateur residents. One wonders if less visually interesting but more possibly damaging problems such as inter-est rates on debt repayment would get the same attention is there were a drone-shaped solution. The use of drones in Africa has also highlighted internal tensions. As Andrews reports, the military use of drones by Amer-ica in the Horn of Africa and the Middle East makes some nervous about flying eyes over their country. Further, governments with secrets to hide would rather not have cameras accidentally gathering embarrassing foot-age – and that might include officials corruptly involved in the poaching business. "Like pretty much every other legal and moral issue, what hap-pens to poachers, drones and wild animals depends on political and eco-nomic priorities" (Andrews 2014). There are alternatives which introduce the same tensions. Raxter and Young write:

> A proven strategy is to combat poaching with skilled, intelligence-driven anti-poaching units that rely on networks of informants within local com-munities and deployment of trained undercover officers. The goal is to understand every link in the poaching chain: Who kills the animals, and where and when, what routes poachers follow out of the kill zones, who conducts the trade in wildlife parts.
>
> (Raxter and Young 2015)

This solution requires long-term investment and doesn't make for such spectacular publicity images. It does have the benefit of addressing the poaching within its context since such crime exists because people are willing to do it and someone is willing to pay for the proceeds. Morozov has talked about a drive for technology to solve problems without addressing underlying issues and to provide us with information but not necessarily greater understanding:

> If technological fixes are inevitable, and if some forms of solutionism cannot be avoided, let us at least make sure that this solutionism is of the self-reflexive, perhaps even neurotic kind. Only through radical self-doubt can solutionism transcend its inherent limitations.
>
> (Morozov 2013: 352)

Investigative reporting, along with the coverage of disasters, appears to be the two areas where drones can add depth and understanding to stories. Indeed the case studies outlined in this chapter suggest that drones can offer stories which otherwise would not be on the agenda, such as looking at the assets of corrupt politicians or monitoring for environmental problems. The debates around the use of drones to combat poaching are a suitable warning that journalists should not view the technology as a simple solution but one that brings its own set of ethical dilemmas.

Notes

1 See for instance American reporter Nellie Blye exposing mental institutions in the nineteenth century, German reporter Gunter Walraff's work documenting dangerous workplaces in the 1980s and numerous examples in the United Kingdom.
2 Long-running protests over an oil pipeline in North Dakota in 2016 inevitably saw drones deployed by newsgathers and protestors. And just as inevitable were efforts by law enforcement agencies to ban them. See: Arrests for filming drones at pipeline protest http://www.grandforksherald.com/news/north-dakota/4142234-2-face-charges-operating-drones-during-pipeline-protests
3 Such hostility has been experienced by others investigating animal rights issues. See for instance http://boingboing.net/2016/09/13/cops-hassle-filmmakers-flying.html
4 For more detail see http://thebureauinvestigates.com.
5 The story can be found here: http://www.nytimes.com/interactive/2015/10/27/world/greenland-is-melting-away.html
6 Quoted in http://www.suasnews.com/2014/02/somerset-floods-filmed-by-multirotor-for-bbc-panorama/
7 You can see the footage here: http://www.11alive.com/news/local/drone-video-shows-blight-in-atlanta-neighborhoods/190237299
8 You can see some of the footage here: https://www.youtube.com/watch?v=KOxh9dbkNT4
9 You can see footage here: https://vimeo.com/68229603

10 See http://agora.md
11 It is outside the scope of this book to consider the history, rationale and scale of such interventions, but it is an area worthy of closer examination.
12 For details see http://www.greenpeace.org/international/en/news/Blogs/Energydesk
13 For more information see www.conservationdrones.org
14 See also http://www.usnews.com/news/articles/2012/12/07/google-to-fund-anti-poaching-drones-in-asia-africa
15 A good summary of the technological pros and cons of drones combatting poachers can be found here: http://www.bbc.co.uk/news/business-28132521 and there is also a discussion with links on this Save the Rhino website: https://www.savetherhino.org/rhino_info/thorny_issues/the_use_of_drones_in_rhino_conservation//starts//

References

Alexander, B. (2016) Drone video shows blight in Atlanta neighbourhood. *11alive.com*. 13 May 2016. http://www.11alive.com/news/local/drone-video-shows-blight-in-atlanta-neighborhoods/190237299

Andrews, C. (2014) Wildlife monitoring: Should UAV drones be banned? *Engineering and Technology* Vol. 9(7), 14 July 2014, 33–35.

Cubby, B. (2013) Drone will range freely over farms to keep tabs on animal welfare. *The Age*. 31 March 2013.

Culver, K. (2014) From battlefield to newsroom: Ethical implications of drone technology in journalism. *Journal of Mass Media Ethics: Exploring Questions of Media Morality* Vol. 29(1), 52–64.

Edge, A. (2015) 'Old stories, new perspectives': Drone journalism in Africa. *Journalism.co.uk*. 8 June 2015.

Goldberg, D.; Corcoran, M. and Picard, R. G. (2013) Remotely piloted aircraft systems & journalism opportunities and challenges of drones in news gathering. *Reuters Institute for the Study of Journalism*. https://reutersinstitute.politics.ox.ac.uk/fileadmin/documents/Publications/Working_Papers/Remotely_Piloted_Aircraft_and_Journalism. pdf

Gonzalez, L.; Montes, G.; Puig, E.; Johnson, S.; Mengersen, K. and Gaston, K. (2016) Unmanned Aerial Vehicles (UAVs) and artificial intelligence revolutionizing wildlife monitoring and conservation. *Sensors* Vol. 16, 97.

Gutermuth, L. (undated) Starting satellite investigations. *Exposing the Invisible*. https://exposingtheinvisible.org/resources/obtaining-evidence/starting-satellite-investigations

Gynnild, A. (2014) The Robot eye witness. *Digital Journalism* Vol. 2(3), 334–343.

Haner, J. (2015) A drone's vantage point of a melting Greenland. *New York Times*. 26 October 2015. http://www.nytimes.com/2015/10/28/insider/a-drones-vantage-point-of-a-melting-greenland.html?_r=4

House of Lords. (2012) *Future of investigative journalism*. Select Committee on Communications.

Houston, B. (2010) The future of investigative journalism. *Daedalus* Vol. 139(2), Spring 2010, 45–56.

Hunter, M. (2011) *Story-based inquiry: A manual for investigative journalists.* Paris: UNESCO.

Keneally, M (2012) Drone plane spots a river of blood flowing from the back of a Dallas meat packing plant. *Daily Mail.* 24 January 2012. http://www.daily-mail.co.uk/news/article-2091159/A-drone-plane-spots-river-blood-flowing-Dallas-meat-packing-plant.html

McChesney, R. (2013) *Digital disconnect.* New York: The New Press.

Morozov, E. (2013) *To save everything click here: Technology, solutionism and the urge to fix problems that don't exist.* London: Allen Lane.

Platt, J. (2016) Drone uncovers illegal logging in critical Monarch butterfly reserve. *Takepart.com.* 22 June 2016. http://www.takepart.com/article/2016/06/22/drones-uncover-illegal-logging-monarch-butterfly-habitat?cmpid=organic-share-twitter

Raxter, P. and Young, R. (2015) Drones can curb poaching but they are much costlier than alternatives. *National Geographic.* 23 May 2015. http://voices.national geographic.com/2015/05/23/drones-can-curb-poaching-but-theyre-much-cost lier-than-alternatives/

Rouse, M. (2016) Showing corruption from the skies. *Internews.* 8 June 2016. https://medium.com/local-voices-global-change/showing-corruption-from-the-skies-4fe34e89c93b#.ulxvpo1w8

Schiffrin, A. (2014) *Global Muckracking: 100 years of investigative journalism around the world.* New York: The New Press.

Tremayne, M. and Clark, A. (2014) New perspectives from the sky. *Digital Journalism* Vol. 2(2), 232–246.

Wallis, P. (2016) Op-ed: Tribe in Guyana makes drone to monitor illegal logging, mining. *Digital Journal.* 18 April 2016. http://www.digitaljournal.com/news/environment/op-ed-tribe-in-guyana-makes-drone-to-monitor-illegal-logging-mining/article/463210#ixzz4MNGF4B2r

Youngs, I. (2014) Wildlife film-makers reveal tricks of the trade. *BBC online.* 24 July 2014. http://www.bbc.co.uk/news/entertainment-arts-28405412

6 Drones in the classroom

The high school students seem to having a great time getting to grips with drones for the first time. In a large hall one student is practicing flying while others look on, until there is a momentary loss of control and a student has to take emergency action to avoid getting hit.[1] It is a reminder that, unlike technology such as mobile phones, this is a skill which brings with it not just questions about privacy but also fundamental concerns about safety. The school is in Holmes County, Florida. In 2015 it started running a course in drones which was in part sponsored by Embry-Riddle Aeronautical University. That university has very close links with private and state aviation industries and also offers what it claims is the first master's in the United States in unmanned and autonomous systems engineering.[2] The high school students earn college credits and the teachers get to engage with them on subjects which they might otherwise find unpopular. Aerospace Career Academy Regional Director Tim Hester said: "Well, the STEM: science, technology, engineer, and math. It's a, it's a cool way to slip in the science and math. The kids don't really realize what they're, how they're engaged" (quoted in Hill 2015). One student told the local television reporter:

> They're not just for surveillance to spy on people. Instead of paying a crop duster to go and do a whole field, you could pay a hundred just for them to go down the field with a drone and spray the few plants that are infected.
>
> (quoted in Hill 2015)

In financial terms the focus of drones in education remains at universities with research through aeronautical centres. However, as the example above shows, the technology is also spreading into other areas.[3] In the United Kingdom, a proposed Environmental Science A-Level will include a section on how drones can be used to monitor poachers, track water sources or survey crops. Richard Genn, who is leading the development of the new award due to be offered from September 2017, said:

> Environmental Science is a rapidly developing subject, and technology plays a crucial role. This new course is designed to be relevant and topical so that new issues can be studied as soon as they emerge – what is on the news today could be studied in the classroom tomorrow.
>
> (quoted in Espinoza 2016)

Broadly drone studies can be broken down into the teaching and research on aeronautics and studies on applications. As discussed earlier, and highlighted by the content for the courses above, the specific use as a media tool is a relatively small part. Setting aside the huge amounts of money put in by defence contractors, civilian applications are concentrated on construction and agriculture, and that is where a lot of the teaching and research focus lies. A survey of American universities involved in drones highlighted Idaho State University looking at detecting blight in potato crops,[4] the University of Pittsburgh teaming up with two others to look at monitoring animal disease vectors and Texas A&M's Centre for Robotic-Assisted Search and Rescue looking at how to incorporate drones. The sole 'media' example was the University of Colorado, which had won funding to look at using drones to collect sound (Chester 2016). As the author says: "The sampling above offers only a glimpse at what's possible in terms of drones advancing research and practice in higher education. Currently, however, the regulatory hurdles challenge rather than facilitate many institutions' ability to pursue these possibilities" (Chester 2016). Looking at the experiences of teachers who have tried to introduce drones into the classroom, there remain structural hurdles to getting the most from the technology. This is particularly pronounced with journalistic applications, which do not have a huge fund of industry research sitting behind them. Instead, the interest and dedication of individual staff members have often driven the teaching.[5]

A good example is Jeff Ducharme at the College of the North Atlantic in Canada. Ducharme was a print journalist for many years before moving into teaching. He saw the work that was going on at the University of Nebraska's Drone Journalism Lab and started experimenting with his own drone. It convinced him that this was something his journalism students needed to be exposed to. "I was very, very fortunate. The then president of the college [Dr. Ann Marie Vaughan] was very big on innovation and as soon as I mentioned it to her she loved it," he said (author interview, February 2016). Getting the equipment and curriculum sorted was very much a labour of love for Ducharme and the small pocket of drone enthusiasts at the college. There was not a manual on how to do this and issues around ethics were not something that departments researching wing design had needed to confront. Indeed it is only in 2016 that the first handbook on teaching drones was published and that is more aimed at high schools.[6] The experimentation outside of the classroom has not been matched by the development of sustainable or tested procedures for teaching it inside the classroom.[7] The research environment has a much stronger base but its practices and expectations are far different. While an increasing number of programmes will touch upon drones through warfare, politics and cultural studies, these tend to be at an abstract level. The experience at the College of the North Atlantic, and the few other such venues, is then very useful though it is still in its infancy. The college began teaching drone journalism in 2013 as part of Photojournalism and added an elective drone journalism course in 2015.[8] Ducharme has made use of the environment around his campus to experiment with storytelling with drones for instance. His campus is in a very beautiful environment so not every student will be able to gather footage flying over a 100-year-old former ice breaker beached nearby.[9] Nonetheless, the principle of using what is around you to help students learn how to construct stories is one which journalism schools around the world do on a daily basis. Interestingly, Ducharme says they spend the least amount of time flying because of the preparation work that needs to go in to understanding the principles, processes and responsibility. Indeed it can be such an exhausting experience, even when they operate in two-person teams, that some find it is not for them. This is a reaction other educators have also noted. "Drones can be a hundred times more stressful than people think," says Ducharme. Ultimately though he believes it is a tool which journalism schools need to be aware of as much as any other newsgathering item.

The experience of journalism staff at the College of the North Atlantic is one that appears to be repeated elsewhere. Matthew Schroyer said:

> Journalism students aren't yet widely exposed to the drone as a tool of journalism in many institutions. It seems to be viewed as a niche area of study, rather than the versatile, capable companion that should be taught alongside DSLRs and HD cameras. Academic institutions also would do well to try to push the boundaries of this technology and find best practices of how the drone can be used in environmental and data journalism projects. That's not to say there hasn't been great work done by many academics, institutions, or other organizations. Just to name a few, the Drone Journalism Lab at the University of Nebraska – Lincoln, the University of Missouri, the College of the North Atlantic, and Michigan State University have all done great work, and new programmes are emerging all the time. The Tow Center for Digital Journalism at the Columbia Journalism School has been doing an excellent

job keeping track of emerging technologies in general and finding out
how they can be applied to journalism.

(author interview, April 2016)

The University of Nebraska has been one of the leading establishments
looking at how drones can be taught as a newsgathering device. Its Drone
Journalism Lab was established in November 2011 under the auspices of Prof
Matt Waite to explore how drones could be used for reporting. "In the lab, students and faculty will build drone platforms, use them in the field and research
the ethical, legal and regulatory issues involved in using pilotless aircraft to
do journalism."[10] A regular collaborator is the University of Missouri, which
has a Drone Journalism Programme run under the auspices of Bill Allen, who
has 25 years' experience as a reporter, and Rick Shaw.[11] The equivalent in
the United Kingdom is the University of Central Lancashire's Civic Drone
Centre, which is leading the way on drone journalism applications along with
other civilian uses such as rescue services.[12] The speed with which drone
research is taking off, and the fledgling nature of drone journalism in particular, is evidenced by the fact that it was only launched in February 2015. While
there is a decade of experience in the aerospace industry, only very recently
has that linked up in further and higher education with media departments.
As well as research, the centre is creating Civil Aviation Authority-approved
drone training and is drafting operating policy of all drone use for the university. Postgraduate drone modules as well as short courses are also being
developed. The centre has also run DroneHack hackathons, where members
of the public and academics look at humanitarian and civic challenges. The
centre expects to expand its teaching and research activities in the coming
years. Prof Paul Egglestone, a former TV producer who oversees the centre,
sees the use of drones remaining as a relatively specialist area: "It's another
tool in the box. I don't think we are going to approach the time when everyone
carries a drone just in case, but there will certainly be specialist camera crews
that acquire a drone for themselves" (quoted in Hartley 2015).

At the University of the West of England in Bristol there is similarly a
long tradition of aerospace research in a city with history of aviation innovation. The prestigious Bristol Robotics Laboratory, run with the University of Bristol, builds on that looking, among many other things, at aerial
robotics.[13] A particular area of research is designing autonomous drones that
could explore inside collapsed buildings. As with other universities, interest
and research comes from many different areas. Dr Mark Palmer is a senior lecture in computer science and creative technologies and also a keen
hobbyist and member of a model flying club.[14] His experience highlights
some of the basic problems of getting students to connect with drones in
the classroom as opposed to in a research laboratory. For one thing there is

getting the right insurance which varies depending on hobby, development and research or that needed for commercial work. Then he recommends there is qualified pilot supervising and ready to take control so that there is always a back-up experienced pilot able to step in. In terms of kit he suggests starting with very small drones which can cause less damage and be easily repaired. You also need to consider battery life since, if you are just getting ten minutes flying time, then for a room of students that might not be enough to get to grips with the controls. Palmer now has signs warning people that filming is in operation which he can deploy as well as having cones and warning tape and, if a risk assessment deemed it necessary, might have additional crew to deal on hand. He said:

> unfortunately for those thinking that drone technology offers an easy fix, everyone flying them needs to consider CAA [Civil Aviation Authority] rules and those seeking to do it commercially need to pass theory and flight exams by an approved trainer as well as submitting an operations manual to the CAA. We have somebody at the club who films for the BBC. When I first met him at the club and mentioned that I was interested in filming sites near us called 'starfish' sites which were fake cities used in World War Two to decoy bombers away from Bristol he started to list the things I needed to consider and having looked into it further I can see why he did this.
>
> (author interview, July 2015)

At the other end of the city, but still part of the university, Peter Venn runs the prestigious MA in wildlife film-making. This is delivered in partnership with the BBC's renowned Natural History Unit. It is a sector which has embraced the possibility of drones for improving their productions. As Venn explains:

> In the last few years, relatively cheap and readily available drone technology has developed rapidly alongside the evolution of increasingly lightweight, miniature cameras and stabilization. This has revolutionised the potential for filmmakers to capture high-quality, yet extremely economical aerial footage. Specifically, in terms of the visual ambitions of student filmmakers, drones have allowed them to film with a blue-chip inspired aesthetic, previously only seen in high-end productions. A few years ago, steady aerial footage, offering good detail in a controlled shot, was the exclusive preserve of expensive wildlife films, which could afford powerful lenses, in stabilized systems all mounted onto helicopters. However, even these bigger budget, wildlife productions have embraced drone technology, because the drone has augmented existing aerial filming technologies largely due to their small size, fine control and maneuverability. In this respect they are

unmatched as they can be flown into places that film crews could never get a helicopter. Moreover, operate as filming platforms at much lower altitude than helicopters, working close to the ground, with little noise and therefore causing minimal disturbance to wildlife. Essentially, the drone has allowed a vast range of wildlife filmmakers from amateur to student and professionals to capture not only impressive footage, but also very revealing film material. Drones are helping to give viewers a far greater appreciation of many wild places simply by providing a fresh perspective where species and habitat are seen in wider context.

(author interview, September 2016)

For the wildlife course students have been experimenting with their own drones and testing out processes in the wild. The experience at UWE with interest coming from many different disciplines as well as staff's personal research is a common theme.

Several academic institutions in the United Kingdom have begun offering short courses aimed at getting people capable of flying at an amateur level but not part of any particular curriculum. For instance Sheffield College runs a two-day course costing £300 aimed at people wanting to film. Prof Steve Thomas, who has worked at the University of South Wales for more than 30 years, set up one-day drone introduction courses. It has had around 35 people paying £100 since they started in the spring of 2015, with a cohort of 10 on each run using a pool of some 15 machines owned by the university. "They do theory in the morning and then simulations. Then they go to the sports centre and fly a drone," he said. "It is about awareness, none of these machines are toys. You have to know how to operate them. People have no idea about the legal requirements. The batteries are like small bombs" (author interview, February 2016). He said that several people had attended looking at how they might use for professional/media purposes but generally, once they realized the legal and other restrictions, they decided it was actually too complex to take on. To get CAA approval for commercial operations (which includes education) involves passing a theory exam and writing an operations manual as well as passing a flight exam. The exam takes place out of doors and involves operations at long distances and height to cover the sorts of operations that might be carried out under CAA rules. "Drones are going to be massive but it is the training, that is the key thing," he said. Prof Thomas has a proposal to set up a drone academy at the university offering a much fuller programme of courses. They have had interest from utility companies who would like to get people trained up. "We have had the police ring us up and ask if they need CAA [Civil Aviation Authority] approval."

So how might drones for journalists in education be developing? An indication can be gleaned from a survey carried out by the author.[15] Academics from 10 UK and 4 overseas educational institutions completed a questionnaire asking about their use of drones for teaching media students and how they thought it might develop over the coming three years. Eight of the 10 said that it was likely or very likely they would increase their teaching of drones over the next three years and half thought their drone fleet would increase in size. However only three said they were likely to have a module dedicated to drone journalism and most did not expect to offer a specific drone qualification or think it likely that they would recruit someone solely on their drone skills. The implication is that drones would still be a part of general journalism teaching, and a growing part, rather than warrant a standalone offer. There was more likelihood of offering a short continuing professional development course which suggests an economic decision on what might attract potential students. The practical and legal difficulties that still come with using drones are likely to be a factor in this approach. Four of the 10 respondents said their department owned their own drones and 7 of the respondents said that staff were not considering getting a Civil Aviation Authority licence to fly. It is more common that another department owns them or that staff have their own. This is backed by anecdotal evidence from academics the author has spoken to. For instance at Nottingham Trent they borrowed one from the Geography Department while at the University of Bristol it was a member of staff who used it as a hobby who supplied the expertise. For the overseas institutions, apart from Whitman College, they were all far more positive about how they might use drones in the future. RMIT in Melbourne, Australia, College of the North Atlantic in Canada and Riphah International University in Islamabad in Pakistan all expected to see an increase in teaching and the ownership of drones with more staff qualified to use them. It still remained very unlikely though that someone would be recruited specifically because of their drone expertise. They also expected to attract research funding in this area, which was felt to be much less likely in the UK institutions. For the United Kingdom, then, it feels that in general drone interest is not yet strong enough in media departments to warrant the standalone provision but as part of existing programmes. One British academic spoke about their experience of introducing drones to undergraduate journalism students. It highlighted an interesting point that sometimes it can be the lecturers who are more interested and motivated than the students:

> So far, it's worked for me to invite students to indicate their interest and out of about 50 media students, literally only about three have wanted to come out and try it, which is manageable. I'm surprised in the lack of interest, but then our media course is more academic than practical. Using a big green campus and applying a common-sense approach to the regulations, we have filmed with no problem. So for me, it works best as an added bonus to my modules; i.e. simply as a

way to get interesting shots to include in their assessment. I include one guest lecture. Not well attended. But I'm interested in thinking about how to include drones more into the theory/academic side of my teaching too.

(author interview, August 2016)

When the House of Lords' European Union Committee reported in 2015 on the civilian use of drones in the European Union it covered a huge range of areas – but barely universities. Committee members visited Cranfield University to look at its pioneering research and teaching on drones, including how they are being deployed for investigations into emergency incidents (House of Lords 2015). Drone research in monetary terms at universities remains dominated by the defence and security sector. Yet there has been a huge increase in civilian research and media responses to drone use, both from a practical and theoretical perspective.[16] The community of practice among journalists sharing ideas has been crucial to developing the use of drones in spite of, rather than thanks to, official regulations which have been confused at best. Sharing of good practice in education has been lagging behind and yet there are still many crucial questions which need to be answered about drone use by the media. Some of these are technical questions about, for instance, safety and some of them are questions about privacy and data collection. The role of education appears to be viewed mostly as providing technological solutions to problems when it can also provide the social context and understanding for how this technology is being deployed. The House of Lords felt it important to recommend that: "UK media regulators should initiate a public consultation on the appropriate use of RPAS by the media, with a view to providing clear guidance." Universities have been at the forefront of developing such guidance, often in conjunction with specific media outlets. There is a nucleus of fascinating work at a small number of institutions which requires better support so that the next generation of drone pilots are schooled in a practice with a strong ethical and practical approach.

Notes

1 Footage can be seen here: https://www.youtube.com/watch?v=FtfvOhuMA9A
2 For details see https://www.erau.edu/degrees/masters-unmanned-autonomous-systems-engineering/index.html
3 An example is the Clarksdale school board in Mississippi which heard about a drone club run by a student and initially it considered banning the drones but then changed its mind and started using them for teaching. http://www.edweek.org/ew/articles/2015/12/23/clarksdale-school-board-oks-drones-as_ap.html
4 Which makes sense since Idaho is famous as a state for its potatoes. See http://idahopotatomuseum.com/
5 http://www.expouav.com/news/latest/drones-in-classroom/

 6 *Drones in Education* by Carnahan, Zieger and Crowley. It describes itself as: "written for educators who want to incorporate drones into their curriculum but have no idea where to start." For more information see http://www.techlearning.com/blogentry/10614?platform=hootsuite
 7 There is a useful discussion here on teaching drones in the classroom which hints at some of the many issues involved: http://er.educause.edu/articles/2016/7/~/link.aspx?_id=BA644F1504534983BC95178C95314E58&_z=z
 8 For more details see http://www.cna.nl.ca/news/News-Article.aspx?messageid=939
 9 See https://www.youtube.com/watch?v=L09ePOTSfLA&feature=youtu.be
10 For more details see http://www.dronejournalismlab.org/about
11 For more details see http://www.missouridronejournalism.com/
12 For more details see http://www.uclan.ac.uk/research/explore/groups/civic-drone-centre.php
13 For more details see http://www.brl.ac.uk/researchthemes/aerialrobots.aspx
14 Declaration of interest – the same institution the author teaches at.
15 The author conducted a survey using Survey Monkey which sent invites via social media for academics to participate. Academics from the following UK institutions responded: Ulster University; Nottingham Trent; Swansea; Strathclyde, Birmingham City; Sheffield; Ravensbourne; University of the West of England; University of Central Lancashire and Northampton. Academics from the following overseas institutions responded: Royal Melbourne Institute of Technology in Australia, Whitman College in Washington State, America; North Atlantic College in Canada and Riphah International University in Pakistan. It should be noted that this survey relied upon people responding and therefore it cannot be considered a sample sufficient from which to make predictions about the actual teaching of drone journalism in further and higher education. Only those who were aware of the social media call could respond and only those interested took part, and they only responded to the best of their knowledge, so it cannot be claimed that it is a definitive survey of all the drone activity at each of these institutions. They qualitative survey sought to gather impressions from those teaching to suggest areas of interest and development in this field.
16 The Centre for Study of Drones at Bard College in America is one example of this community of practice with this weekly drone round-up email offering a comprehensive oversight.

References

Chester, J. (2016) Earl days in drone use in higher education. *Educause Review*. 11 July 2016. http://er.educause.edu/articles/2016/7/early-days-for-drone-use-in-higher-education

Espinoza, J. (2016) Teens to study use of drones in new A-level. *The Telegraph*. 30 June 2016. http://www.telegraph.co.uk/education/2016/06/29/teens-to-study-use-of-drones-in-new-a-level/

Hartley, S. (2015) The rise and rise of drone journalism. *Contributoria*. May 2015. http://www.contributoria.com/issue/2015–05/550c2bda24f8eb5f32000219/

Hill, B. (2015) Drones hover over Holmes District High School. *News Channel 7*. 6 March 2015. http://www.wjhg.com/home/headlines/Holmes-District-Students – 295403001.html

House of Lords. (2015) *Civilian use of drones*. European Union Committee, 7th report of session 2014–2015.

7 Navigating ethics and regulations

When a gas explosion blew apart a house in Grimsby in the north of England in 2015 it exposed a new front in the battle between the right to publish and the right to privacy. IPSO was established in September 2014 as one of two rival press regulators in the United Kingdom following the abolition of the Press Complaints Commission (PCC). IPSO was set up by publishers as an alternative to Impress, which met the recommendations the government's Leveson inquiry into press standards. The bulk of the United Kingdom's newspapers signed up to IPSO. It carried on, as the PCC had, in using the industry's own Editor's Code of Practice as its guide on how to behave. A little over a year after it began work, IPSO was asked to make its first adjudication on an alleged breach of privacy involving a drone.

The complainant was upset because of footage carried in her local paper, *The Grimsby Telegraph*, of her house which had been ripped apart by a gas explosion. Susan House complained that the drone flew over her garden, the footage showed the contents of her home which were not visible to members of the public, including her bathroom, stairs and bedroom. Ms House also said that she was occupying the property at the time, and that no one expects a drone to fly over their property taking videos.[1] The Code of Practice is put together by a committee and was updated on 1 January 2016.[2] It covers areas such as reporting stories involving children, discrimination, intrusion into grief and privacy. The latter is covered by Section 2, which says that everyone is entitled to privacy in their private home, family and health. Specifically it says: "It is unacceptable to photograph individuals, without their consent, in public or private places where there is a reasonable expectation of privacy." Therefore editors will be expected to justify such intrusions on the grounds of public interest – and half a dozen such grounds are listed. *The Grimsby Telegraph* argued that the footage was taken at a distance such that no personal belongings or individuals were depicted. It said that no one was living at the property as it was extremely badly damaged.

And its public-interest defence was that showing the amount of damage that can be caused by a gas explosion helped people think about gas safety.

Three months after Ms House submitted her complaint IPSO came down on the side of the newspaper. It said that while Ms House's bath and shower could be seen in the footage little else could be discerned and it agreed there was public interest in showing the damage. Not only that: "Because of the extent of the damage, it would not have been possible to do so without showing some of what had previously been the internal contents of the house."[3]

While IPSO used the word drone in their adjudication, its Code of Practice has not been updated to take account of the new technology. A spokesman for IPSO said:

> The Editors' Code was reviewed and reissued in January [2016] with no specific mention of drones. However, as to whether we are satisfied that the Code allows us to apply existing principles (particularly in Clauses 2 and 3) to new technological developments in newsgathering, I would say that we are confident that the rules let us consider complaints about drones. The Clauses are broad enough to permit us to take drone use into account and the House judgements are a good example of that.
>
> (author interview, April 2016)

In this chapter I will look at the regulatory and ethical issues surrounding the use of drones for newsgathering. As the example above shows, the public, law enforcement, regulators and the media are already clashing over drones while cheaper technology and popularity mean their use is becoming more widespread. The rules governing their use by the media are already becoming an unwieldy patchwork, with some organizations quite advanced in what they require of their employees, while others have barely considered them, and laws on flying, data gathering and privacy vary enormously. This chapter will move between different countries and I acknowledge that not all comparisons are equal. First amendment free speech rights in America are a lot stronger than in the United Kingdom, while people in the United Kingdom are generally more accepting of surveillance (e.g. with CCTV cameras) than in other European countries. Nonetheless there are some general approaches which journalists in these different countries share, even if their validity has been critiqued. These include the idea of journalists as a watchdog upon the institutions of society, independent of partisan interests and the government and free to investigate, comment and criticise. The Code of Practice used by IPSO, for instance, is founded on this approach. If the operation of drones is to be analyzed then it needs to be within the context of the codes journalists define themselves by. However, while IPSO believes its code is sufficient not to require any changes to take account of drones, other media organizations disagree. There is a debate over how much drones alter the power of the media and therefore what regulations are required. According to Culver, the technology can "fundamentally alter" existing privacy protection:

> In the same way telephone lenses extended the perimeter from which photos could be taken, drones can alter the space from which images and data can feasibly be captured. For instance, a person has no reasonable expectation of privacy on a public beach but many would object to the practice of a news organization using a drone to capture an ongoing all-day livestream of a beach for constant broadcast. The calculation becomes not a legal one – private versus public property – but instead an ethical one – considering a person's desire to be left alone, even in shared spaces.
>
> (Culver 2014: 59)

There is also another driver, of less interest to journalists but as important, and that is the pressure by commercial organizations to have a regulatory system which best allows them to do business. The ethical concerns for protecting individuals may yet come off second best.

Impress, the rival press regulator to IPSO, has also adopted the Editor's Code of Practice as its guide, though it is consulting on adopting a new set of standards.[4] Meanwhile broadcasters have been quick to update their guidance. This may be because there is stronger regulatory oversight in the United Kingdom of broadcasters through Ofcom and also that regulations covering newspapers leave it to them to agree on operational policies such as health and safety. The Channel 4 guidelines on drone use run to more than 1,000 words and are hosted in the general section and the secret filming section.[5] All production crew are expected to read the guidance. If there is to be an invasion of privacy then there needs to be a strong public-interest reason and a written application needs to go to a Channel 4 lawyer. One production company which successfully navigated these rules was Windfall Films, which produced the 'Hidden Britain by Drone' programme. Producer Joe Myerscough said it took "many hours of discussion" with the channels' lawyers before they got the go-ahead. The biggest issue was privacy and Channel 4 approached the programme in the same way it would as if it was using covert filming. In other words the justification and the extent of footage used had to meet the highest threshold. Drones are hardly covert, certainly at low level since they are not silent, but it is clear there are sensitivities over what they might record. According to Myerscough the trickiest places to film both legally and logistically were in built-up areas:

> One of the most difficult sequences was Bow Quarter in East London. This used to be the infamous Bryant and May match factory where the match girls strike of 1888 took place and now it has been turned into a secure gated community, cut off from the rest of the city with its own swimming pool and supermarket. Despite all being private land, it was an incredibly busy place with people constantly coming and going which kept us grounded for long periods. We had a huge team of marshals controlling the exits and footpaths – if you watch the sequence in the programme you can see all our marshals pretending to look at their phones so they blend into the background. We also had to reassure the residents that we weren't snooping through their windows.
>
> (author interview, July 2016)

Other broadcast organizations have guidelines in place. In 2014, CNN released material aimed at educating journalists and private citizens on the proper uses of drones (Holton et al. 2015: 644).

The BBC also has guidance on the use of drones for its employees which was last updated, at the time of writing, in June 2016.[6] The guidance sits alongside policies on privacy, safety and data protection, and there are 14 main points covering essential rules such as the pilot needs to be properly qualified and authority must be sought from a senior executive. As with Channel 4, any use where there is an invasion of privacy and consent has not been obtained from the subject must be treated the same as covert filming; and therefore the bar is set high on justifying the intrusion. There are several points worth looking at in more detail. One is that the BBC says before user-generated content (USG) from drones is aired, journalists need to consider if any regulations were broken in gathering the material:

> Where the BBC is offered user generated content and it appears the drone flight put the safety of people or property at risk or has otherwise been carried out illegally, including a breach of aviation regulations, any use of the footage gathered must be justified in the public interest.

With the growing number of people using drones and a clear market for their footage, it seems an eminently sensible policy to have in place so that journalists know how to judge whether to use such material. As with mobile phone images, the capacity to record information places a lot of power in the hands of citizens but they may not have the understanding of the ethical and legal issues at play. The BBC also says:

> Drones should not normally be used to identify individuals without their consent, or capture close-up images of private areas such as houses, gardens or offices without the consent of the owner, unless these areas can be seen from a public vantage point or there is a public interest in showing them.

That phrase 'public interest' was also used in the UGC policy and means that, while the threshold is set high, the possibility of intrusion exists if the wider public interest in terms of exposing wrongdoing, criminal behaviour or dangers to the public is served. The final part of the guidelines I want to draw attention to says:

> Drone operators should normally comply with the laws and regulations that apply to the airspace where the drone is flown, unless there is a strong editorial justification for not doing so. In a war zone it may not be possible to seek permission to fly. In the absence of country-specific drone regulations, operators should aim to follow the UK CAA standards.

Again, there is this allowance for the guidance to be ignored if the story is strong enough to support it. This is a well-established policy for the media across different platforms but it does rely on the institution being robust enough to be able to critically reflect on what it is doing and decide not to go ahead if necessary. The ethical codes by which they operate, aside from any potential legal ramifications, therefore need to be very strong because once intrusion has taken place it cannot be undone. And, as the BBC acknowledges, abroad or where civil society has broken down, there is the greatest need to have a strong ethical code to operate by because there is not any legal sanction to hold the journalist back. If it is in a disaster area where the pressure might be on to get footage back as soon as possible and everyone is concentrating on dealing with victims, the idea of referring back to CAA rules might be overlooked. This is a challenge for journalists and their media organizations – and even more so for those freelancers who have even less incentive to accept restrictions on what they gather. As we shall see later in this chapter, and as the very presence of the BBC guidelines demonstrate, there is actually quite a lot of thinking by journalists about how drones should be used.

The organization which regulates how personal data is collected and shared in the United Kingdom is the Information Commissioner's Office (ICO) and it has added a specific section on its website about using drones. As it says, if a drone has a camera, then it has the potential to be covered by the Data Protection Act. In particular the ICO advises people to think before sharing images gathered on social media and for keeping those images safe.[7] For those using them professionally it points them towards guidance issued as part of CCTV and surveillance. 'Responsible' journalistic activity does have an exemption under the Data Protection Act and journalists would not necessarily think about themselves or their camera or their notebook as a surveillance tool with its pejorative overtones. Here the ICO starts from that perspective. Section 7.3 of this guidance doesn't equivocate when it comes to warning about the potential for misuse of what it calls Unmanned Aerial Systems (UAS):

> The use of UAS have a high potential for collateral intrusion by recording images of individuals unnecessarily and therefore can be highly privacy intrusive . . . as such, it is very important that you can provide a strong justification for their use.[8]

The ICO highlights, as with other surveillance technology, the fact that people being observed need to be able to identify who is doing the observing. "You will need to come up with innovative ways of providing this information," the ICO says. Furthermore, the ICO says:

> For example, this could involve wearing highly visible clothing identifying yourself as the UAS operator, placing signage in the area you are operating UAS explaining its use and having a privacy notice on a website that you can direct people to, or some other form of privacy notice, so they can access further information.

It would be interesting to see how many media outlets have adopted such measures.

In many discussions on the use of drones for journalism, the issue raised more than safety is privacy. "The use of drones in newsgathering raises a number of intriguing issues for journalists . . . one issue of importance is that of privacy rights" (Tremayne and Clarke 2014: 235). However the reason for this concern varies. A report from PwC on the future of the drone industry looked at what barriers there are to growth in this sector. After safety the next main concern was around privacy, especially on how to handle the "vast amounts of data, sometimes including confidential or sensitive information about private property or private behaviour". The report says:

> Market growth increases the pressure to regulate this area, though it will take time to prepare and pass proper legislation. This shouldn't be a major factor preventing further adoption of drone technologies, just as it wasn't in the case of telecommunication, internet and mobile technologies.
>
> (PwC 2016: 24)

Here the concern appears to be simply that regulation is required to ensure the market can operate effectively. There appears to be no particular ethical framework on which to base such legislation other than that, whatever it is, it enables companies to expand. Consider similarly a report put together by an expert panel convened by the University of Birmingham looking generally at what the United Kingdom needed to do to use of drones for military and policing purposes but which also covered civilian use. In an interesting choice of language, the report says: "We note in passing that lightweight RPA surveillance is also likely to become the weapon of choice for paparazzi in search of intimate photographs of celebrities" (Birmingham Policy Commission 2014: 76). It recommended that by 2035: "There will be well understood and, effectively enforced, restrictions on all private RPA use to protect privacy. A media complaints system under the Royal Charter will adjudicate on paparazzi intrusions" (Birmingham Policy Commission 2014: 9). This suggests a very narrow understanding of how the media use drones with protection given to celebrities ahead of anyone else. However, this concern about unregulated photographers is a regular part of the discourse when it comes to discussing how the media might abuse drones. In Clarke's analysis, the civilian use of drones, the use by journalists, is dominated by what he calls "a form of debased or corrupted journalism" which is essentially tabloid journalism. Clarke warns: "Given the dedication of paparazzi, and the money that can be made from 'scoop' pictures of celebrities and notorieties, it is readily predictable that there will be frequent abuses of the power of drone-borne cameras" (Clarke 2014: 240). The wedding of singer Tina Turner in Switzerland, a trip by singer Beyoncé on a rollercoaster and numerous Californian celebrities simply at home have attracted paparazzi drones (Sheridan and Graham 2014). In 2015, Californian governor Jerry Brown signed into law a much tougher definition on what constituted a "physical invasion of privacy" to cover drones taking video or photos above peoples' homes. This was aimed directly at countering the activities of paparazzi (Megerian 2015). While the public interest in seeing a celebrity by their garden pool is negligible, other cases are less clear cut. Dutch paparazzi Rene Oudshoorn used a drone to get an image of a disgraced banker which sparked a national debate about privacy. Oudshoorn said:

> Van Keulen didn't want to talk to journalists, so I went there and flew over his house. I was lucky, because he was in his backyard. While flying, security came outside. They called the police and despite my permit, I was taken to the bureau. That also became part of the front page story.
>
> (quoted in Postema 2015: 74)

Such concerns about paparazzi are not new. In 1888 Kodak produced its first camera. It was a portable box which took photos on a roll of film rather than plates. As Barber and Wickstead note:

> At five guineas, the Kodak was far from cheap, but it made practical photography more widely accessible than ever before. It ushered in an era of similar cameras, whose small size, portability and ease of use allowed more surreptitious photography than was previously possible. These cameras were often talked of generally, and some were marketed specifically, as 'detective cameras'.
>
> (Barber and Wickstead 2010: 246)

Matt Waite from the University of Nebraska's Drone Journalism Lab has likened the reaction to drones to that which accompanied the advent of the Kodak camera:

> Signs were posted barring Kodak Brownies from beaches because people were afraid to be photographed in their bath suits. Scholars and lawyers wrote long essays about how public life was destroyed forever because strangers with small cameras could record everyone's private moments.
>
> (quoted in Carroll 2015: 21)

In 1890, Warren and Brandeis wrote an influential paper for the *Harvard Law Review* entitled "The Right to Privacy". As they put it:

> Recent inventions and business methods call attention to the next step which must be taken for the protection of the person, and for securing to the individual what Judge Cooley calls the right 'to be left alone'. Instantaneous photographs and newspaper enterprise; and numerous

mechanical devices threaten to make good the prediction that 'what is whispered in the closet shall be proclaimed from the house-tops'.

(Warren and Brandeis 1890: 195)

The concerns of these jurists were sparked by photographers sneakily obtaining pictures of actresses using this new technology – and it is a discourse which we see repeated today. In his analysis of this paper and the possible development of privacy law in regard to drones, Calo says that such law tends to respond to specific incidences or abuses.[9] The general increase in surveillance (by the state, by corporations or by fellow citizens) and how much people are willing to accept has been accompanied by only piecemeal legal protection; a pattern repeated in many countries. What Calo calls a process of acclimation. As Deuze said: "As is common with media in general and surveillance systems in particular, we tend only to become acutely aware of them when they break down" (Deuze 2012: 117). The fear shared by both corporations and drone enthusiasts is that an individual egregious wrongdoing will spark legislation which will be hard to undo and have far-reaching consequences for the 'law abiding' user. According to Calo the data-gathering abilities of drones and their potential to be a ubiquitous presence thanks to civil and police use makes them a game-changer when it comes to privacy. "As with previous emerging technologies, advocates will argue drones threaten our dwindling individual and collective privacy. But unlike the debates of recent decades, I think these arguments will gain serious traction among courts, regulators and the general public" (Calo 2011: 32). The reason is that they literally open up a new front on what might be regarded as private and that they can hoover such huge amounts of information.

As well as this fear of the paparazzi, Margot Kaminski, assistant professor of Law at Ohio State University, has identified another powerful narrative about privacy and drones – the sunbathing teenager. Kaminski highlights stories from Kentucky, New Jersey, Virginia, Connecticut, California and Florida which all follow the same basic pattern: complaints that a drone was monitoring ordinary members of the public often while they were on a beach or in their garden and often these happened to be young women (Kaminski 2016). The story has crossed the Atlantic and there is a report of a sunbathing woman in Bristol forced to quickly cover up after a drone flies overhead. Kaminski believes the appeal of the story has its roots in the story of Lady Godiva and Peeping Tom punished for his lust. Since such stories will generally be illustrated by sunbathing women, there may be other reasons editors find it useful to run. For Kaminski, though, the key problem is that it distracts from real issues about drones and privacy and it is the same one that Calo highlights – the potential for data gathering and who regulates this. For Kaminski, technologies such as facial recognition and the ability to cross check information with other databases make these a potentially powerful tool for intrusion and one which civil law hasn't addressed. She writes:

> As social actors, we regularly use cues from our physical and social environments to decide how much we want to disclose in a particular setting. Technologies like drones disrupt environments. They take down walls. They distance the human operator from the enforcement of social norms in a particular setting (like the beach). They disrupt our ability to calculate how much we've disclosed by potentially tracking our behaviour over time, at far lower cost than a helicopter.
>
> (Kaminski 2016)

The data-gathering potential for drones means that it is not just the target that should be considered. In my car my phone connects to the stereo via a short-range Bluetooth link. The page on the phone shows all the potential Bluetooth connections at any one time. As I drive down the motorway it is electronically waving, if not connecting, with hundreds of machines. We emit a huge amount of electronic information as a matter of routine and a drone is another source for gathering such data – should it be so configured. Even if not directed to capture such information, there is the potential for collateral intrusion. A drone flying past a building may accidentally record people several stories up in their home or place of work. If a camera was filming in a street in the United Kingdom gathering general views, then people would not have much say about being recorded. If they were in a building and on the twentieth floor, they would not expect to be filmed. Indeed they may not even know they had been filmed because the sound of the drone wouldn't be heard through the window. If they did know, how would they complain? What information would they have to know who was operating the drone and why? The operator is 20 floors below and could well be gone by the time the person makes it to the ground floor. Drones are not big enough to carry very visible logos of whichever particular broadcaster is operating them. Therefore the review and storage of such footage would need to be considered by the media operator. An article in Slate in 2016 did a rudimentary test of what a drone might see flying next to a building. The implication was that, with better technology such as facial recognition or high-resolution cameras, it was hard not to gather data on people; though the sound of the drone was often a giveaway (Ismail 2016). Much as Google cars recorded everything for the Streetview and then had to bow to pressure to blur people and places,[10] one should be wary of seeing the pattern repeated for drones. As Serafinelli writes:

> Previous surveillance studies emphasised the widespread general passive acceptance of the violation of privacy (given by the Internet and CCTV, for instance). Drone technology augments that phenomenon introducing the element of mobility, which represents an additional manifestation of the McLuhanian theorisation of media as extensions of man.
>
> (Serafinelli 2016)

Technological solutions may well be presented. For instance drones might broadcast their details electronically so that their ownership can be traced. One already in place for some commercial drones is geofencing, whereby the GPS system has details on particular areas, such as airports, which are off limits and therefore the drone is prevented from flying there even if the operator requests it. There is research into various active measures to defend areas by 'hacking' drones if they stray into protected areas, essentially an electronic counter-measure. The danger is that any hacking might be too aggressive and cause the drone to malfunction and therefore become a safety risk. Writing in the *Guardian* in April 2015, Dr Jonathan Aitken, research fellow in robotics at the University of Sheffield, warned that without a robust regulatory environment the potential for the industry to take off could be stymied. Aitken favoured licencing saying:

> However, a clear distinction needs to be drawn between a simple drone and a toy. This could be based on weight: drones over 4kg or a self-built drone over 1kg should be registered and require a flying licence that links a pilot to a drone, which should then be stored onboard. While regulation will provide a deterrent, it does not provide comprehensive protection as many operators are unaware, or are willingly ignoring regulation, as the spate of recent arrests for drone incursion has shown.
>
> (Aitken 2015)

The idea of geofencing (that is, having electronic no-go areas) did not, for Dr Aitken, offer foolproof protection. As someone who researches autonomous systems, his solution was more research into autonomous systems:

> Research should be directed not just to understand what a vehicle can do but to determine what a vehicle will never do, for example not entering a geo-fenced area, even under human control. By reliably

understanding these boundaries of operation we can impose useful ground rules that will keep us safe around these vehicles.

(Aitken 2015)

These technical fixes do not, however, address the issue about whether such intrusion should take place – they only accept that it will and try to offer mitigation. It becomes then a technological race rather than an ethical debate. Rao et al. summarise the discussion:

> Typically, as in other emerging technology-based products, the customers want choice and access, businesses want to manufacture and sell products unhindered while meeting customer needs, and the regulatory and law enforcement agencies need to have confidence in the regulations themselves as well as their ability to enforce violations.

(Rao et al. 2016: 87)

The ethical dimension appears to be absent in the pursuit of technological advance and profit.

Industry is certainly doing its best to get its version of what it thinks makes good regulations heard. In 2016 there was a call from aviation associations in Europe which issued a joint letter requesting a registration policy. A total of 10 associations in the European Union made the call, including the European Cockpit Association and the International Air Transport Association, though not the United Kingdom's Civil Aviation Authority. The 10 said that the ability to trace nuisance drones back to their owner or pilot would improve compliance with regulations.[11] As of 31 March 2016, in the 16 largest economies in the world, drone registration is required in just six of them; Argentina, China, Indonesia, Russia, Turkey and America.[12] However there are wide variations about what else is required to undertake a commercial flight. The International Civil Aviation Organization (ICAO), which is an arm of the United Nations, has 191 members, of which 63 have some regulations for drones already in place; nine states have pending regulations and five have temporarily banned the use of drones.[13] This call for co-ordinated responses is regularly repeated. In April 2014 Siim Kallas, the European Commission vice president responsible for transport, said there should be tightened regulations on drones:

> Many people, including myself, have concerns about the safety, security and privacy issues relating to these devices. That's why we are proposing to draw up a range of tough measures to regulate this market, while at the same time allowing European industry to take advantage of the opportunities it presents.

He quoted figures that the drone industry in Europe could be worth 15 billion Euros a year by 2024, clearly a market worth protecting.[14] In June 2016, commercial drone manufacturer DJI said that governments in Europe needed to work together and collectively send data on no-fly zones and flight restrictions to a centralized database so that companies could better geo-fence their machines. The company's European policy lead, Christian Struwe, said: "We encourage governments not to regulate simply out of fear or due to sensationalised news coverage. We need better, more proactive communications between governments and us, the drone industry" (Russon 2016). One other way that there could be better communication is in keeping the public informed about their privacy rights. The University of Central Lancashire's (UCLAN) Civic Drone Centre produced a response to a government consultation on drone development. One of UCLAN's identified barriers was: "Privacy & Data Protection – better awareness for the public that they are protected by existing regulations and laws" (Ansell and Heaton 2015: 3). Meanwhile a House of Lords report on drones said: "UK media regulators should initiate a public consultation on the appropriate use of RPAS by the media, with a view to providing clear guidance " (House of Lords 2015: 52).[15]

Currently the situation is confusing for the media as well as the public. As Schroyer points out, with specific reference to American law:

> While some aspects of the regulatory environment seem to be improving, others seem to be getting worse. States, counties, and municipalities are working to create their own laws and ordinances that are harmful to the practice of drone journalism. In Texas, for example, it's illegal to use a drone to photograph private property without permission, even when the private property is in plain public view.
>
> (author interview, April 2016)

Research by Fifield in 2016 found that:

> Twelve states placed new restrictions on drone users this year, with two of the laws, in Oregon and Vermont, restricting law enforcement or government use. Other new laws address recreational drone use, which is becoming more common. A new law in Kansas makes it illegal to stalk with a drone. In Arizona, Louisiana and Utah, drone operators are now restricted from flying drones near police or firefighter activity. And in Oklahoma and Tennessee operators can't fly drones near some buildings, such as power plants.
>
> (Fifield 2016)

University of Nebraska professor Matt Waite said: "There are legislators that are going to look for feel-good solutions that are going to make legitimate things illegal, and they are going to infringe upon First Amendment rights" (quoted in Fifield 2016). Other commentators have weighed in with similar complaints.[16] This obviously focusses on the situation within America and, since these reports, the FAA has implemented new rules on commercial flying of drones which bring some clarity to the situation for journalists. Nonetheless this confusion of regulations is repeated elsewhere. In the United Kingdom, as we have seen, guiding journalists using drones is the Civil Aviation Authority, whatever regulatory body looks after your sector (Ofcom for broadcasters for instance), the Office on the Information Commissioner, your employer's rules (if relevant or if they even exist) and then how, on the ground, people actually interpret the rules. People who teach courses to pass Civil Aviation Authority exams so they can become drone pilots say there are plenty of areas where the issue isn't black and white. As television producer Joe Myerscough found out when he met for hours with broadcasting lawyers to get the go-ahead on his drone programme:

> They sought the advice of lots of external experts but we soon learnt that the grey areas weren't all that grey, or they were very dark or very light shades of grey that could easily have been confused with black and white.
>
> (author interview, July 2016)

One effect of this lack of knowledge or clarity around regulations is that media workers are at risk when they use drones. An analysis of cease-and-desist letters sent by the American Federal Aviation Authority to drone operators found that 13 out of 17 were for use of aerial photography or video for commercial purposes. Two universities also received such letters for using drones for "journalism education purposes". In another example a television station was warned that it could be in violation of FAA rules after it aired footage of damage by a tornado taken by one of its employees using their privately owned drone (Holton et al. 2015: 640). Meanwhile in the United Kingdom, photojournalist Eddie Mitchell working with CAA approval was arrested by the police in 2014 after they appeared spooked that he was flying near Gatwick airport.[17] He subsequently said he would sue for wrongful arrest and said he had carefully followed all the rules.[18] In February 2015, three journalists working for the television station Al-Jazeera were arrested for a flying a drone on the outskirts of Paris.[19] The city was on a heightened state of terror alert although, again, the trio appeared to have followed regulations with one flying the drone, another operating the camera and a third observing. As Schroyer points out:

> Police [in America] have been granted permission to shut down large swaths of airspace in order to deny news helicopters access to critical events, as was the case during the political unrest in Ferguson, Missouri, in 2014. In Hartford, Connecticut, police questioned a photographer at a crime scene for using a drone, and put pressure on a local news station to fire the photographer. So the challenges aren't just what kind of certification you have or where you can fly your equipment, it's also about the real challenges you will face day-to-day on the ground.
>
> (author interview, April 2016)

Koebler suggests: "If you do find yourself in legal trouble from the FAA or state or local law enforcement, knowing the law helps, but it's not necessarily going to get you out of trouble" (Koebler 2016).

For all these many valid concerns, professional drone users have proactively embraced the idea that they should try and codify their behaviour. Indeed it is remarkable how much effort has gone in to drawing up guidance just by those involved in the media, environmental or humanitarian efforts. We saw at the beginning of the chapter how broadcasters in the United Kingdom had written relatively comprehensive policies. In May 2016 there was a recommendation to have a code of practice for people using drones to monitor wildlife. Jarrod Hodgson from the University of Adelaide's Unmanned Research Aircraft Facility said that while drones were a worthwhile ecological survey tool for scientists, limited research had been undertaken into how their use may affect animals.[20]Meanwhile an extensive set of guidelines have been produced by UAViators on how drones should be used responsibly in humanitarian settings.[21] Directly applicable to journalists, Jeff Ducharme, a former print reporter turned lecturer at the College of the North Atlantic in Canada, came up with drone journalism code in 2014.[22] He teaches journalism students how they might use drones and through this process developed the code. The 21 points are clearly written by someone who has used drones extensively and thought carefully about their use – these are not generic regulations for photographers with just the nouns switched. For instance, as many who have flown drones testify, it can be a draining business and therefore the more difficult the environment, the better it is to have at least two people working so that the roles are split. This kind of knowledge only comes through practical application. The majority of the code is about safe flying and the onus is on safety over any public interest in the story. Point four says: "A drone is a powerful tool and it must be treated as such. A drone should only be used to gather information pertinent to a given story. Drones should not be used to search for stories." Ducharme is explicit that the code is not just for journalists but also for members of the public. "I put the code together because we need to tell the public that we can operate safely and ethically," he said (author interview, February 2016). "This is just the starting point for a discussion as I am sure it will develop but I needed to put a framework together. I see the ramifications if we don't; we could lose this" (author interview, February 2016). Ducharme's starting point for his code was the Professional Society of Drone Journalists (PSDJ), which has its own ethical guidance.[23] The PSDJ says that its code should be read as another layer alongside other codes journalists might have: "Additionally, such considerations should not be viewed as the sum total or end of the ethical requirements for drone journalists, but is the minimum that is expected of a drone journalist." Developed by Matthew Schroyer, he sets out a hierarchy rather than just a bullet point of rules. It starts with newsworthiness ("The investigation must be of sufficient journalistic importance to risk using a potentially harmful aerial vehicle. Do not use a drone if the information can be gathered by other, safer means"), and then moves though safety, sanctity of the law and public spaces, privacy and finally traditional journalistic ethics. Rather than a list of rules, Schroyer says he hopes this will allow a more flexible approach to judging individual situations. Then in the summer of 2016, after the FAA announced new rules on how the media could use drones, the Drones Journalism Lab released an operations manual for reporters.[24] Put together by Matt Waite and Ben Kreimer, it has checklists for people to follow to ensure safe flying and, near the beginning, a section on ethics. Drone journalists are asked to: "Balance the public's need for information against potential harm or discomfort. Pursuit of the news is not a license for arrogance or undue intrusiveness." General rules on appropriate behaviour are given as well as directing people to the Society for Professional Journalists. As with the BBC rules highlighted at the beginning of the chapter, if there is sufficient public interest then a journalist may breach someone's privacy, but the language of the codes says this is very much an exception and other legal and ethical guides should be taken into account. It offers a far more nuanced approached to newsgathering than some commentators who see journalism as either about paparazzi or everything else.

There is always going to be a tension between regulators and the media over the use of drones for newsgathering. The codes which journalists work by, and which generally they have written themselves, share common attributes and one is that crudely, and safety aside, intrusion is justified if the public interest is strong enough. Regulators, whether for aviation or data, are not motivated by the same fourth estate concerns. Corporations, and the media is made up of corporations, are looking for regulations which enable them to operate as efficiently as possible. Somewhere among all this are citizens – and they may also be flying drones and not subject to any of the codes and rules the others are judged by. As Peter Lee, from the legal firm Taylor Vinters, told the House of Lords, the Civil Aviation Authority needed to facilitate access by bona fide journalists to breaking news events because: "If this particular developing area of rapid response journalism by RPAS is ignored then irresponsible, amateur cameramen will, in all likelihood, attempt to take footage anyway" (House of Lords 2015: 52). In America at least there appears to have been a careful and deliberate effort to come up with ethical guidelines so that drones can be used by reporters in a safe but useful manner. The test will come when a serious accident or incursion takes place involving drone journalists and how the state and the public react. A cursory glance at the history of the last 50 years shows that there are regular crisis points where the media comes into conflict with legal, state or economic interests.[25] That is unlikely to end and drones will get their flashpoint. Photographer Lewis Whyld was an early adopter of drones and has used them both in the United Kingdom and abroad. He said:

> The need in this country [the United Kingdom] is not as great as it is elsewhere – both in terms of commercial and journalistic uses. My UAVs are primarily designed for use in other countries. I plan on invading the privacy of dictators, despots and human rights abusers who massacre civilians with impunity. Every generation of journalists has a responsibility to use the tools available to them in order to discover and report the truth.
>
> (author interview, June 2012)

Notes

1 Ms House made six complaints in total and these can be found here: https://www.ipso.co.uk/IPSO/rulings/IPSOrulings.html
2 A copy of the code can be found here: https://www.ipso.co.uk/IPSO/cop.html
3 A copy of the ruling can be found here: https://www.ipso.co.uk/IPSO/rulings/IPSOrulings-detail.html?id=337
4 See http://impress.press/standards/
5 You can see a copy of the regulations here: http://www.channel4.com/producers-handbook/c4-guidelines/secret-filming-guidelines
6 A copy of the guidance is available here: http://downloads.bbc.co.uk/rmhttp/guidelines/editorialguidelines/pdfs/Drones-guidance.pdf
7 Details can be found here: https://ico.org.uk/for-the-public/drones/

8 You can see more detail on this section here: https://ico.org.uk/media/1542/cctv-code-of-practice.pdf

9 Calo's amusing example is that no provider of videos can release customer rental history because journalists once obtained a list of videos enjoyed by a supreme court nominee (Calo 2011: 29).

10 Google introduced this automatic policy in 2008: https://www.cnet.com/uk/news/google-begins-blurring-faces-in-street-view/

11 Details on the request can be found here: http://www.bbc.co.uk/news/technology-37285825?ocid=socialflow_twitter

12 The situation is very fluid however. At the end of 2016 Sweden decided to treat drones like CCTV camera making their use far more difficult (see: http://www.thelocal.se/20161220/sweden-set-to-reverse-controversial-drone-ban) while the UK's transport minister suggested a 'driving test' for all drone operators (See https://www.theguardian.com/technology/2016/dec/21/drone-users-face-safety-test-new-uk-regulations)

13 Information from PwC report.

14 Quoted in this press release: http://europa.eu/rapid/press-release_STATEMENT-14–110_en.htm?locale=en

15 See also this news report on the House of Lords findings, which adds useful commentary to its findings: http://www.dronejournalism.org/news/2015/3/house-of-lords-looks-to-media-regulators-for-guidance-on-drone-journalism

16 See for instance this column in *The Economist*: "The current state of the law – both legislation and court decisions – is poorly suited to deal with persistent surveillance. This is because privacy law is tailored to questions of whether one is in public – an open field – or in a space where one has a 'reasonable expectation of privacy'. The Supreme Court has, at times, expanded such spaces, for instance finding in 1967 that the FBI cannot eavesdrop on conversations in telephone booths without a warrant. But in this era of 'big data', the line between public and private can no longer be delimited by physical boundaries." More here: http://www.economist.com/blogs/democracyinamerica/2015/03/drones-and-privacy?fsrc=scn/tw/te/bl/aloomingthreat

17 See https://www.theguardian.com/uk-news/2014/dec/31/drone-photojournalist-arrested-gatwick-aiport-near

18 See http://www.pressgazette.co.uk/drone-photographer-sue-police-false-imprisonment-after-arrest-scene-surrey-caravan-fire

19 See https://www.yahoo.com/news/3-al-jazeera-journalists-arrested-paris-flying-drone-180511717.html?ref=gs

20 See http://www.abc.net.au/news/2016–05–24/animals-need-protection-from-research-drones/7439218

21 See http://drones.frd.ch

22 You can download a copy from here: http://www.academica.ca/top-ten/cna-prof-develops-drone-journalism-code-conduct

23 You can see it here: http://www.dronejournalism.org/code-of-ethics

24 You can download a copy here: http://www.dronejournalismlab.org/manual/

25 The Pentagon Papers, Thalidomide Scandal, Cash for Questions and Snowden leaks are just some of the examples where efforts have been made to thwart the exposure of wrongdoing. This is aside from structural problems because the political economy of media corporations makes their primary purpose the pursuit of profit.

References

Aitken, J. (2015) We need ground rules if drone technology is to take off. *The Guardian.* 7 April 2015. http://www.theguardian.com/media-network/2015/apr/07/drone-technology-effective-regulation-law

Ansell, D. and Heaton, A. (2015) *Responses to questions asked by BIS ahead of challenge business programme – UAV workshop 23/11/2015.* University of Central Lancashire.

Birmingham Policy Commission. (2014) *The security impact of drones: Challenges and opportunities for the UK.* Birmingham: Birmingham Policy Commission.

Calo, M. (2011) The drone as privacy catalyst. *Stanford Law Review Online* Vol. 64, 29.

Carroll, R. (2015) *God's eye news: The use of drones in journalism, a documentary film.* Illinois State University thesis, paper 427.

Clarke, R. (2014) Understanding the drone epidemic. *Computer Law & Security Review* Vol. 30, 230–246.

Culver, K. (2014) From battlefield to newsroom: Ethical implications of drone technology in journalism. *Journal of Mass Media Ethics: Exploring Questions of Media Morality* Vol. 29(1), 52–64.

Deuze, M. (2012) *Media life.* Cambridge: Polity.

Fifield, J. (2016) How drones raised privacy concerns across cyberspace. *PBS.* 1 July 2016. http://www.pbs.org/newshour/rundown/how-drones-raised-privacy-concerns-across-cyberspace/

Holton, A.; Lawson, S. and Love, C. (2015) Unmanned aerial vehicles. *Journalism Practice* Vol. 9(5), 634–650.

House of Lords. (2015) *Civilian use of drones in the European Union.* European Union Committee. 5 March 2015.

Ismail, A. (2016) What can consumer drones actually see? *Slate.* 12 May 2016. http://www.slate.com/articles/technology/future_tense/2016/05/how_much_can_consumer_drones_actually_see.html

Kaminski, M. (2016) Enough with the 'sunbathing teenager' gambit. *Slate.* 17 May 2016.

Megerian, C. (2015) Gov. Jerry Brown approves new limits on paparazzi drones. *Los Angeles Times.* 6 October 2015. http://www.latimes.com/local/political/la-pol-sac-brown-drones-paparazzi-20151006-story.html

Postema, S. (2015) *News drones: An auxiliary perspective.* Edinburgh Napier University thesis. May 2015.

PwC. (2016) Clarity from above: PwC global report on the commercial applications of drone technology. *PwC.* May 2016.

Rao, B.; Gopi, A. and Maione, R. (2016) The societal impact of commercial drones. *Technology in Society* Vol. 45, 83–90.

Russon, M. (2016) DJI: Drone industry really needs central database of flight restrictions across Europe. *International Business Times.* 9 June 2016. http://www.ibtimes.co.uk/dji-drone-industry-really-needs-centralised-database-flight-restrictions-across-europe-1564637

Serafinelli, L. (2016) *Drone technology and visual ownership: Privacy and Security issues.* Paper given at the IAMCR.

Sheridan, P. and Graham, C. (2014) Attack of the drones: Hollywood celebrities are besieged by paparazzi spies in the sky. Worried? You should be . . . because they'll soon be a regular fixture over YOUR home. *Mail on Sunday.* 6 September 2014. http://www.dailymail.co.uk/news/article-2746231/Attack-drones-Hollywood-celebrities-besieged-paparazzi-spies-sky-Worried-You-ll-soon-regular-fixture-YOUR-home.html#ixzz4MXnofoZX

Tremayne, M. and Clark, A. (2014) New perspectives from the sky. *Digital Journalism* Vol. 2(2), 232–246.

Warren, S. and Brandeis, L. (1890) The right to privacy. *Harvard Law Review* Vol. 4(5), 15 December 1890, 193–220.

8 Over the horizon

In April 2016 Facebook CEO Mark Zuckerberg revealed the company's 10-year road map for how it will develop. It was looking at artificial intelligence, virtual reality and improved connectivity, the latter delivered in part by drones. As part of the slick presentation which no technology company can do without, a drone from partner company DJI appeared. "At Zuckerberg's prompting, a drone flew onstage and streamed the surprised crowd live to Facebook" (Strange 2016). The Facebook CEO told the audience: "We are building the technology to give anyone the power to share anything they want with anyone else" (D'Onfro 2016). Zuckerberg's stunt does highlight some of the tensions at play as media companies start to experiment with drones. The idea that a multi-billion dollar company gives anything away without a sound business reason is questionable. The description of an empowering technology which can liberate people is familiar. In Jackman's illuminating study on drone trade shows, he records "a persistent and insistent framing of the drone as inevitable: an unrelenting force of possibilities, which any regulatory tide cannot and ultimately will not quell" (Jackman 2016: 3). The fact that Facebook, whose founder has previously said that social media means people do not have an expectation of privacy (Johnson 2010),[1] is building technology involving drones should give us pause for thought. One might assume that the audience for Zuckerberg's talk were glad to participate in his spectacle. Perhaps by attending they implicitly gave consent to be part of the show and for that to be streamed live and recorded and stored somewhere. The report by Strange says they were surprised, which suggests prior consent had not been sought. If they did not, what power would they have to say no?

This issue of surveillance is, I believe, a key one as to how drones will develop because it highlights the influence not just of those make the technology but those who use it and those who are the subject of it and the relative powers at play. According to Klauser and Pedrozo, 'surveillance' is not a suitable term, as it implies routine and systematic information gathering while media use of drones is "sporadic and punctual". The authors write: "Approaching drones as techniques of vision and visualization, rather than as techniques of surveillance, allows a wider focus that remains open to the unsystematic visibilities created by drones, as increasingly mass-marketed devices in manifold public and private use" (Klauser and Pedrozo 2015: 287). Nonetheless, I would argue the common perception is of drones as surveillance devices. What is fluid is who records, who consumes and who can influence the data gathered. Rather than the traditional panopticon approach of Foucault, civilian drones may be used by corporate interests to produce and share news stories, by law enforcement to carry out their duties or by independent users for play or professional use. At times it is more like a *synopticon*, where the many watch the few (Mathiesen 1997); at others, *sousveillance*, where the powerful are watched by the powerless or what Kelsey calls *civeillance*, where civilians watch each other (Kelsey 2015). As Deuze writes: "Surveillance has been slowly but surely moving away from the all-seeing eye of the state and the police to the much more widespread and distributed gaze of the private many" (Deuze 2012: 126).

All of these different users would be answerable in varying degrees to some ethical and regulatory framework, though the degree to which they observe this might vary from group to group, or even from situation to situation. For instance, a strong perceived public interest need may encourage a journalist to use a drone in a situation where normally they would not. The subject of the drone's attention may be unaware by accident – the drone is flying too high to hear – or by design – it is recording covertly. In that case the subject will have difficulty influencing how the content is used. Overt newsgathering with drones, much as media workers make themselves known at events wearing tabards or just because of the kit they are carrying, means that subjects have some ability to negotiate on how they are surveilled. Limited though this might be, it is a long way from Foucault's all-seeing and unaccountable prison warden. The subjects in the examples in this book have included celebrities, farmers, prison guards, corrupt politicians, disaster workers, shanty town dwellers and refugees. Clearly that is only a selection of all drone use by the news media. However, they are all actors in legitimate news stories. There has been no general 'fishing' for stories. In some cases, how they have been treated has raised ethical concerns. The intrusion into the wedding of a celebrity, for instance, raises a question about stories in the public interest or stories which interest the public, but the mere fact it was happening is certainly a story. The ability to affect that coverage varies wildly. California has passed a state law preventing paparazzi using drones, presumably after lobbying by well-heeled citizens, but there is little influence that refugees can bring to bear. This is a very pronounced example of the asymmetric use of this technology. The refugees can't engage with the camera; indeed in many cases it is not clear they even know the camera is above them. There can be no expectation of privacy on their part but there is also little ability to influence how they are portrayed. These are vulnerable people reduced to a crowd, moving in an abstract landscape and given no autonomy over their actions. The Magnum photographer Robert Capa said that if your pictures were not good enough it was because you were not close enough. The drone, circling high above, has reduced the object of its gaze both literally and metaphorically. As Lyon asserts "consumer capitalism continually innovates in the quest for new markets" but:

> the tendency is for the asymmetrical relationship between corporate organisation and individual consumer to be exaggerated by every new gadget and service. Claims regarding consumer benefit – some of which may be perfectly legitimate – hardly have time to be tested before the next innovation appears.
>
> (Lyon 1994: 150)

Yet this asymmetric relationship is not fixed. Indeed the media is alive with stories of people bringing drones down and often expressing a righteous feeling of victory over nosy neighbours[2] or, like John Connor in the 'Terminator' films, using native ingenuity to scupper high-tech snooping.[3] Much as a common response to Google Glass wearers from those being observed was to announce they did not want this imposed surveillance, there is a resistance to these flying eyes.[4] And just as the subjects of drone vision do not always passively accept their role, so we should not expect all users to operate in predictable ways. As Akrich says: "Designers thus define actors with specific taste, competences, motives, aspirations, political prejudices, and the rest and they assume that morality, technology, science and economy will evolve in particular ways" (Akrich in Bijker and Law 1992: 208). However, as Akrich goes on to demonstrate, though designers may inscribe their vision on an object, the user (or actor) can subvert its intended use. Even sealed 'black box' technology may be used, amended or retooled.[5] Drones have from the beginning attracted hacking and modifications and therefore their development is not necessarily in the hands of manufacturers. Adam Rothstein warns against viewing the drone as either one of two robot narratives – the mindless slave which enables us to do great things or the monster which gets out of control:

> What drones will end up becoming is dependent on how the drone meme changes, and it will necessarily differ from the memes that came

before it as the technology on which it is based collides and separates in a chaotic swarm. This could be the technology that writes a new story, as it takes a new place in our world.

(Rothstein 2015: 145)

Similarly Deuze calls upon us to not passively accept the gaze of the drone. "It is in the ways we choose to act under such conditions of many and multiplying all-seeing mediated eyes that we can take responsibility for our actions and thus for the kind of society we want to live in" (Deuze 2012: 128).

In economic terms, the media use of drones will be dwarfed by that in other industries such as agriculture and construction (let alone the military). However the drone on its programmed path surveying acres of crops to detect if any are suffering from disease will not impact upon the public's imagination half as much as the one recording the results of a road accident for the local news. At the moment, the discourse in the media when discussing civilian drones appears to veer between panics about threats to bring down aircraft, aid criminals or help voyeurs and uncritical awe in the face of technological fantasies. This often immature and narrow discussion contrasts with the very thoughtful reflections of many media drone practitioners. There is a growing community of practice crossing continents with people sharing ideas and experiences backed by varying ethical codes. Some of the most interesting work is taking place not in America or the United Kingdom but in Africa, Asia or countries with calcified political systems. The fact that ethics is an integral part of most learning environments for drone journalism is vital since they are increasingly deployed in areas where civil society has broken down. That is when there is the strongest need for an ethical code on how drones are used because there is so little to check upon a reporter. This contrasts with a confused mix of responses by media regulators and aviation authorities. The direction of drone use by the media is being driven by activists, enthusiasts, teachers and early adopters. That is not to say that corporate media is not taking an interest – indeed we would be surprised if it did not – and that interest is taking drone use in many different directions.[6] In 2016 the University of Florida hosted a drone race. The 10-yard dash doesn't sound too onerous – however these were controlled by competitors' brain waves. The feasibility of this had been demonstrated back in 2013 and the US Department of Defense is funding research into this area (Dearen 2016). Drone racing is now a multi-million dollar business. For one commercial manufacturer the idea is to make the object ubiquitous at work and play. Michael Perry, DJI's director for strategic partnerships, said: "Your personal smartphone is a little bit of both. You use it for taking pictures of your family as much as responding to emails from work. We think that over time, our Phantom series will function like that" (quoted in Stayton 2016). As Kellner writes: "Media culture itself proliferates ever more technologically sophisticated spectacle to seize audiences and increase the media's power and profit" (Kellner 2003: 1). Drones fit very neatly into this business plan.

There are structural attributes to journalism that drones will not necessarily solve or bypass. These attributes include a decline in civic reporting, a reliance on official sources, the amount of press release material recycled among diminishing outlets (Davies 2008, McChensey 2013), the narrow social background of key journalists (The Sutton Trust 2006, Turvill 2014), the political economy of news organizations (Herman and Chomsky 2002) or the news values which underpin reporting (Harcup and O'Neill 2016). A drone will not cover a council meeting and it will not quickly rewrite a press release. In the drive towards automative reporting, it might be programmed to fly over sports events and demonstrations but it can't tell you the individual stories of those taking part or the context.[7] Though as the public become aware of how the media harvest news, one could see in the future demonstrations where messages are directed vertically as well as horizontally – the demands not just on placards and hashtags but bite-size stories pinged out to the hovering drone for relaying. The drone, though, cannot question why it is being asked to look at celebrity weddings more than demonstrations against injustice. A drone requires an extra set of skills and it needs to be operated with care and attention – it isn't like adding an SEO-friendly headline as you write the news story. It can provide new images to your celebrity coverage and it is another source of user-generated content – but if that is all it adds to your operation then it is like giving your reporters mobile phones and only using the text function. Coverage of the migrant movements of 2015 is an example, I would suggest, where drones have added spectacle but not necessarily insight. The pictures of streams of people walking along motorways or across farmland certainly give an idea of the scale of what is happening in that particular place at that specific time. The sweeping shots of a Greek beach littered with detritus left by migrants add a self-consciously filmic dimension to the story that draws you in. It sparks a response one might find similar to that identified by Castro when she analyzed films of battlefield aerial footage shot just after the end of World War One:

> The fluidity of the camera movement acts undeniably as a source of emotion: emotion linked to the pleasure of discovering the earth from a fresh point of view; emotion attached to the sudden recognition of the land as one more wounded body; and emotion, finally, arising from being able to travel freely through space-time. The cinematographic

specificity of this film is fundamental, for no assemblage of aerial photographs could possibly convey, so immediately and effectively, the
cenesthetic intensity by the doubled kineticism of the flight and the
cinematrographic views.

(Teresa Castro in Dorrian and Pousin 2013: 125)

In addition eyewitness testimony has long had a powerful role in journalism, adding authority and establishing credibility (Gynnild 2014, Zelizer
2007). Drones, much as with camera phone footage but often with greater
impact because of the quality of the images, have the effect of placing
the viewer at the scene and they can be supplied not just reports but also
through participants opening up new sources. Indeed they give the viewer
not just an eyewitness view but by their very nature a god-like view. It is an
intoxicating vision. However, in the case of some of the migrant coverage,
it does not get us any closer to the subjects. It doesn't tell us their names
or their personal stories. The god-like view implies omniscience but it can
feel like voyeurism. The drone images themselves are not sufficient to tell
the story in this instance. They are used to draw the viewer in, especially
by placing the images at the beginning of the bulletin, but fundamentally
they add little knowledge. They perform a very functional role of giving an
alternative selection of shots to help construct the package. The decline in
mainstream coverage of key parts of civic society led media commentator
Josh Stearns to write:

We are entering an era of 'hindsight journalism', where some of the
most important stories of our time emerge after the fact. This kind of
journalism shines a spotlight on critical issues, but serves as more of an
autopsy than an antiseptic. It dissects issues like specimens, instead of
shining a light on problems before or as they emerge.

(quoted in McChesney 2013: 181)

Yet, as outlined in this book, there are examples where drones can shine a
light. The journalists in Moldova looking over the wall to count the luxury cars owned by their country's elite. The freelancer avoiding gagging
laws to investigate agricultural malpractice. The economic divide in South
Africa illuminated by pictures juxtaposing shanty towns and golf courses.
A report by the BBC on the future of news raises the possibility of drones
being used to gather data far wider than simply images but to support the
growing trend for sensor journalism (BBC 2015: 28).[8] Meanwhile Stanford
University predicts widespread use of artificial intelligence in drone use by
2030, which could enable media workers to make much better use of their
data-gathering potential.[9] The Federal Aviation Authority rule changes in

2016 are likely to be a significant boost to the media and academic use of drones. It is likely that Europe and the United Kingdom will move towards a similar licencing system.[10] Huge social and economic forces are at play. In 1844, Turner painted *Rain, Steam and Speed*, which depicted a locomotive charging out of the London mist, about to obliterate a hare caught on the line and then hurtle off the edge of the canvass. The unstoppable power of modernity. The revolutionary spectacle from the train was, as Robert Hughes describes,

> a sense of that space which few people had experienced before – the succession and superimposition of views, the unfolding of landscape in flicking motion (the poplars nearby seeming to move faster than the church spire across the field) due to parallax. The view from the train was not the view from the horse. It compressed more motifs into the same time. Conversely, it left less time in which to dwell on any one thing.
>
> (Hughes 1991: 12)

We are not at the mercy of drone technology and there is no preset destination. Instead there is a chance to seize the opportunity to create a multiplicity of drone uses and not any few determined by corporate or state interests. As Clay Shirky wrote in 2009 when the business models for traditional media were collapsing: "Nothing will work, but everything might. Now is the time for experiments, lots and lots of experiments" (quoted in McChesney 2013: 173).

Notes

1 Though reports from June 2016 after Zuckerberg posted a picture to Instagram which showed he had taped over his laptop's camera and microphone suggest he has concerns about his own privacy. See for instance: http://www.independent.co.uk/life-style/gadgets-and-tech/news/mark-zuckerberg-seen-covering-up-his-webcam-in-picture-celebrating-instagram-milestone-a7094896.html

2 The violent responses that drones seem to inspire and the pitfalls around surveillance are studied by Kelsey in one particularly fascinating incident (Kelsey 2015). Meanwhile public opinion surveys in the UK and USA reveal decidedly mixed reactions to civilian drone use. See http://www.theregister.co.uk/2017/01/04/drone_study_mod_dft_public_dialogue/ and http://dronelife.com/2016/12/23/confusing-results-new-survey-public-perception-drones/ and this using drones for deliveries http://www.chicagotribune.com/bluesky/technology/ct-drones-delivery-skeptics-wp-bsi-20161012-story.html

3 See for instance the New Zealand schoolboy who took out a drone by kicking a football at it: https://www.theguardian.com/sport/video/2015/oct/29/new-zealand-boy-knocks-drone-out-of-sky-with-football-video

4 There are numerous reports on reactions to Google Glasses. For BBC reporter Rory Cellan-Jones, who wore them for six weeks, the privacy issue was only

one of a number of factors which meant the product did not become accepted. See http://www.bbc.co.uk/news/technology-27585766

5 A study by Firmino and Trevisan on CCTV operators highlights the autonomy they tried to assert over the technology rather than simply moving a camera around. See Firmino, R. and Trevisan, E. (2012) Eyes of glass: Watching the watchers in the monitoring of public places in Curitiba, Brazil. *Surveillance & Society* Vol. 10(1): 28–41.

6 Interesting examples of where private philanthropy, corporate interest, academic research and citizen activism collide on drones are the annual Drones for Good competition. See this illuminating report from the BBC http://www.bbc.com/future/story/20150910-the-futuristic-city-that-wants-to-rule-from-the-sky?ocid=twfut and this report from a winner http://conservationdrones.org/2015/02/12/data-mule-drone-is-the-national-champion-in-dubais-drones-for-good-competition/

7 Weather reports, stock market stories and the Olympics are some of the story types which have been automated. See for instance: https://techcrunch.com/2016/08/05/robots-will-cover-the-olympics-for-the-washington-post/

8 See also the research done by the Tow Center for Digital Journalism in this area: http://towcenter.org/research/sensors-and-journalism/

9 See https://ai100.stanford.edu/sites/default/files/ai_100_report_0916fnl_single.pdf

10 A House of Lords report from 2015 on the civilian use of drones recommended: "We have already recommended the creation of an online database through which commercial RPAS [Remotely Piloted Air Systems] pilots can provide details of their flights to inform other airspace users. We heard compelling arguments as to why the leisure use of RPAS presents risks to the general public and other airspace users. Therefore, in the long term, we foresee the need for a system which can track and trace all RPAS, especially those flying below 500ft, irrespective of whether they are flown by commercial or leisure pilots. This will be essential not only to manage the increased traffic in the sky, but also to enforce existing and future laws governing RPAS use." The view of many of those in the industry the author spoke to was that a licencing system was inevitable to manage the commercial exploitation of this technology.

References

BBC. (2015) The future of news. *BBC*. 28 January 2015. http://www.bbc.co.uk/news/uk-30999914

Bijker, W. and Law, J., ed. (1992) *Shaping technology/building society: Studies in sociotechnical change.* Boston: MIT.

Dearen, J. (2016) Mind. Blown. Brain-controlled drone race pushes future tech. *AP*. 22 April 2016. http://bigstory.ap.org/d04cc633285c468b8f31f2214cf2feac

Deuze, M. (2012) *Media life.* Cambridge: Polity.

D'Onfro, J. (2016) Facebook just showed us its 10-year road map in one graphic. *Business Insider.* 12 April 2016. http://uk.businessinsider.com/facebook-f8-ten-year-roadmap-2016-4?_ga=1.76197458.672310212.1456410238&utm_content=buffer7359a&utm_medium=social&utm_source=twitter.com&utm_campaign=buffer?r=US&IR=T

Dorrian, M. and Pousin, F., ed. (2013) *Seeing from above: The aerial view in visual culture.* London: IB Taurus.

Gynnild, A. (2014) The Robot eye witness. *Digital Journalism* Vol. 2(3), 334–343.

Harcup, T. and O'Neill, D. (2016) What is news? News values revisited again *Journalism Studies*, DOI: 10.1080/1461670X.2016.1150193, 1–19

Herman, S. and Chomsky, N. (2002) *Manufacturing consent: The political economy of the mass media.* New York: Pantheon.

Hughes, R. (1991) *The shock of the new: Arts and the century of change.* London: Thames and Hudson.

Jackman, A. (2016) Rhetoric of possibility and inevitability in commercial drone tradescapes. *Geographica Helvetica* Vol. 71, 1–6.

Johnson, B. (2010) Privacy no longer a social norm, says Facebook founder. *The Guardian.* 11 January 2010. https://www.theguardian.com/technology/2010/jan/11/facebook-privacy

Kellner, D. (2003) *Media spectacle.* London: Routledge.

Kelsey, D. (2015) Discourse, affect and surveillance: Gender conflict in the omniopticon. *Journalism and Discourse Studies*, November 2015 (issue 2). 1–21.

Klauser, F. and Pedrozo, S. (2015) Power and space in the drone age: A literature review and politico-geographical research agenda. *Geographica Helvetica* Vol. 70, 285–293.

Lyon, D. (1994) *The electronic eye: The rise of surveillance society.* Cambridge: Polity.

Mathiesen, T. (1997) The viewer society: Michel Foucault's panopticon revisited. *Theoretical Criminology* Vol. 1(2), 215–234.

McChesney, R. (2013) *Digital disconnect: How capitalism is turning the Internet against democracy.* New York: The New Press.

Rothstein, A. (2015) *Drone.* London: Bloomsbury Academic.

Stayton, J. (2016) What does the future hold for drones? China may know. *CNN Money.* 16 May 2016. http://money.cnn.com/2016/05/16/technology/drones-future-dji-china/

Strange, A. (2016) Everything you need to know about Facebook's 10-year plan. *Mashable.com.* 12 April 2016. http://mashable.com/2016/04/12/facebook-10-year-plan/#HZt34ZcSjSqu

The Sutton Trust. (2006) The educational background of leading journalists. *The Sutton Trust.* June 2006. http://www.suttontrust.com/wp-content/uploads/2006/06/Journalists-backgrounds-final-report.pdf

Turvill, W. (2014) Study reveals the growing dominance of privately educated elite at the top of UK journalism. *Press Gazette.* 28 August 2014. http://www.pressgazette.co.uk/study-reveals-growing-dominance-privately-educated-elite-top-uk-journalism

Zelizer, B. (2007) 'On having been there': 'Eyewitnessing' as a journalistic key word. *Critical Studies in Media Communication* Vol. 24(5), 408–428.

Index